# Energy Storage: Fundamentals and Applications

# Energy Storage: Fundamentals and Applications

Edited by Anthony Washington

SYRAWOOD
PUBLISHING HOUSE

New York

Published by Syrawood Publishing House,
750 Third Avenue, 9th Floor,
New York, NY 10017, USA
www.syrawoodpublishinghouse.com

**Energy Storage: Fundamentals and Applications**
Edited by Anthony Washington

International Standard Book Number: 978-1-68286-610-8 (Hardback)

### Cataloging-in-Publication Data

Energy storage : fundamentals and applications / edited by Anthony Washington.
    p. cm.
Includes bibliographical references and index.
ISBN 978-1-68286-610-8
1. Energy storage. 2. Force and energy. 3. Power (Mechanics). I. Washington, Anthony.
TJ165 .E54 2018
621.042--dc23

# TABLE OF CONTENTS

# PREFACE

Harnessing energy and storing it in devices like accumulators is known as energy storage. The field concerns itself with converting the energy in a form which is economically suitable to store. The methods involved in this area are thermal storage, mechanical storage, nuclear fuel storage, interseasonal thermal storage, etc. For someone with an interest and eye for detail, this book covers the most significant topics in the field of energy storage. It elucidates new techniques and their applications in a multidisciplinary approach. The textbook aims to serve as a resource guide for students and experts alike and contribute to the growth of the discipline.

To facilitate a deeper understanding of the contents of this book a short introduction of every chapter is written below:

Chapter 1- Energy storage is the harnessing of energy that is stored to be used at a later time. A device that stores energy is called an accumulator. Energy is converted into easily storable forms in this process. This chapter is an overview of the subject matter incorporating all the major aspects of energy storage.

Chapter 2- Accumulator stores, retrieves and releases energy. Various types of accumulators include steam accumulators, rechargeable batteries, flow batteries, etc. The chapter serves as a source to understand the major categories related to accumulators.

Chapter 3- Grid energy storage seeks to store energy in an electrical grid. Compressed air energy storage and cryogenic energy storage are some of the ways to store energy in a grid. The aspects elucidated in this chapter are of vital importance, and provide a better understanding of energy storage.

Chapter 4- Hydrogen storage for later use is usually done by converting hydrogen into liquid hydrogen and slush hydrogen. Liquid hydrogen is stored using cryogenic processes. Compressed hydrogen is stored in tankers and large storage units. The topics discussed in the chapter are of great importance to broaden the existing knowledge on energy storage.

Chapter 5- Power-to-x refers to the excess electricity that is used by energy storage and conversion pathways when energy from renewable sources fluctuates. Energy carrier, Leyden jar, dry-seal Wiggins gasholder, etc. are some of the other topics that have been explored. Energy storage is best understood in confluence with the major topics listed in the following chapter.

I owe the completion of this book to the never-ending support of my family, who supported me throughout the project.

**Editor**

# A Brief Introduction to Energy Storage

Energy storage is the harnessing of energy that is stored to be used at a later time. A device that stores energy is called an accumulator. Energy is converted into easily storable forms in this process. This chapter is an overview of the subject matter incorporating all the major aspects of energy storage.

## Energy Storage

The Llyn Stwlan dam of the Ffestiniog Pumped Storage Scheme in Wales.
The lower power station has four water turbines which can generate a total of 360 MW
of electricity for several hours, an example of artificial energy storage and conversion.

Energy storage is the capture of energy produced at one time for use at a later time. A device that stores energy is sometimes called an accumulator. Energy comes in multiple forms including radiation, chemical, gravitational potential, electrical potential, electricity, elevated temperature, latent heat and kinetic. Energy storage involves converting energy from forms that are difficult to store to more conveniently or economically storable forms. Bulk energy storage is currently dominated by hydroelectric dams, both conventional as well as pumped.

Some technologies provide short-term energy storage, while others can endure for much longer.

A wind-up clock stores potential energy (in this case mechanical, in the spring tension), a rechargeable battery stores readily convertible chemical energy to operate a mobile phone, and a hydroelectric dam stores energy in a reservoir as gravitational potential energy. Fossil fuels such as coal and gasoline store ancient energy derived from sunlight by organisms that later died, became buried and over time were then converted into these fuels. Food (which is made by the same process as fossil fuels) is a form of energy stored in chemical form.

Ice storage tanks store ice frozen by cheaper energy at night to meet peak daytime demand for cooling. The energy isn't stored directly, but the work-product of consuming energy (pumping away heat) is stored, having the equivalent effect on daytime consumption.

# History

## Prehistory

The energy present at the initial formation of the universe is stored in stars such as the Sun, and is used by humans directly (e.g. through solar heating or sun tanning), or indirectly (e.g. by growing crops, consuming photosynthesized plants or conversion into electricity in solar cells).

As a purposeful human activity, energy storage has existed since pre-history, though it was often not explicitly recognized as such. Examples are the storage of dried wood or another source for fire, or preserving edible food or seeds. Another example of mechanical energy storage is the use of logs or boulders in ancient forts — the energy stored in logs or boulders at the top of a fortified hill or wall was used to attack invaders who came within range.

## Recent History

In the twentieth century grid electrical power was largely generated by burning fossil fuel. When less power was required, less fuel was burned. Concerns with air pollution, energy imports and global warming have spawned the growth of renewable energy such as solar and wind power. Wind power is uncontrolled and may be generating at a time when no additional power is needed. Solar power varies with cloud cover and at best is only available during daylight hours, while demand often peaks after sunset. Interest in storing power from these intermittent sources grows as the renewable energy industry begins to generate a larger fraction of overall energy consumption.

Off grid electrical use was a niche market in the twentieth century, but in the twenty first century it has expanded. Portable devices are in use all over the world. Solar panels are now a common sight in the rural settings worldwide. Access to electricity is now a question of economics, not location. Powering transportation without burning fuel, however, remains in development.

## Methods

## Outline

The following list includes natural and other non-commercial types of energy storage (other than in addition to those designed for use in industry and commerce:

- Fossil fuel storage
- Mechanical
  - Compressed air energy storage (CAES)
  - Fireless locomotive
  - Flywheel energy storage
  - Gravitational potential energy (device)
  - Hydraulic accumulator

- o Pumped-storage hydroelectricity (pumped hydroelectric storage, PHS, or pumped storage hydropower, PSH)
- Electrical, electromagnetic
  - o Capacitor
  - o Supercapacitor
  - o Superconducting magnetic energy storage (SMES, also superconducting storage coil)
- Biological
  - o Glycogen
  - o Starch
- Electrochemical (Battery Energy Storage System, BESS)
  - o Flow battery
  - o Rechargeable battery
  - o UltraBattery
- Thermal
  - o Brick storage heater
  - o Cryogenic energy storage
  - o Liquid nitrogen engine
  - o Eutectic system
  - o Ice storage air conditioning
  - o Molten salt storage
  - o Phase Change Material
  - o Seasonal thermal energy storage
  - o Solar pond
  - o Steam accumulator
  - o Thermal energy storage (general)
- Chemical
  - o Biofuels
  - o Hydrated salts

o   Hydrogen storage

o   Hydrogen peroxide

o   Power to gas

o   Vanadium pentoxide

## Fossil and Nuclear Fuel Storage

In today's energy economy dominated by fossil-fuel-based, the storage of fossil fuels (as well as nuclear fuels) is the dominant method to balance the yearly fluctuations between world energy consumption and total primary energy supply (TPES).

## Mechanical Storage

Energy can be stored in water pumped to a higher elevation using pumped storage methods and also by moving solid matter to higher locations. Other commercial mechanical methods include compressing air and flywheels that convert electric energy into kinetic energy and then back again when electrical demand peaks.

## Hydroelectricity

Hydroelectric dams with reservoirs can be operated to provide peak generation at times of peak demand. Water is stored in the reservoir during periods of low demand and released when demand is high. The net effect is similar to pumped storage, but without the pumping loss.

While a hydroelectric dam does not directly store energy from other generating units, it behaves equivalently by lowering output in periods of excess electricity from other sources. In this mode, dams are one of the most efficient forms of energy storage, because only the timing of its generation changes. Hydroelectric turbines have a start-up time on the order of a few minutes.

## Pumped-storage

The Sir Adam Beck Generating Complex at Niagara Falls, Canada, which includes a large pumped storage hydroelectricity reservoir to provide an extra 174 MW of electricity during periods of peak demand.

Worldwide, pumped-storage hydroelectricity (PSH) is the largest-capacity form of active grid energy storage available, and, as of March 2012, the Electric Power Research Institute (EPRI) reports that PSH accounts for more than 99% of bulk storage capacity worldwide, representing around 127,000 MW. PSH reported energy efficiency varies in practice between 70% and 80%, with claims of up to 87%.

At times of low electrical demand, excess generation capacity is used to pump water from a lower source into a higher reservoir. When demand grows, water is released back into a lower reservoir (or waterway or body of water) through a turbine, generating electricity. Reversible turbine-generator assemblies act as both a pump and turbine (usually a Francis turbine design). Nearly all facilities use the height difference between two water bodies. Pure pumped-storage plants shift the water between reservoirs, while the "pump-back" approach is a combination of pumped storage and conventional hydroelectric plants that use natural stream-flow.

## Compressed Air

A compressed air locomotive used inside a mine between 1928 and 1961

Compressed air energy storage (CAES) uses surplus energy to compress air for subsequent electricity generation. Small scale systems have long been used in such applications as propulsion of mine locomotives. The compressed air is stored in an underground reservoir.

Compression of air creates heat; the air is warmer after compression. Expansion requires heat. If no extra heat is added, the air will be much colder after expansion. If the heat generated during compression can be stored and used during expansion, efficiency improves considerably. A CAES system can deal with the heat in three ways. Air storage can be adiabatic, diabatic, or isothermal. Another approach uses compressed air to power vehicles.

## Flywheel Energy Storage

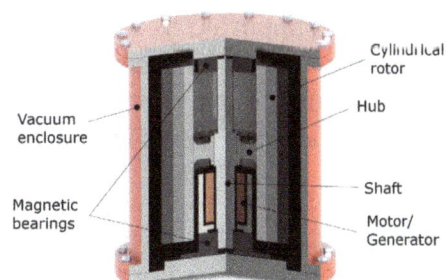

The main components of a typical flywheel

Flywheel energy storage (FES) works by accelerating a rotor (flywheel) to a very high speed, holding energy as rotational energy. When energy is extracted, the flywheel's rotational speed declines as a consequence of conservation of energy; adding energy correspondingly results in an increase in the speed of the flywheel.

Most FES systems use electricity to accelerate and decelerate the flywheel, but devices that directly use mechanical energy are under consideration.

FES systems have rotors made of high strength carbon-fiber composites, suspended by magnetic bearings and spinning at speeds from 20,000 to over 50,000 rpm in a vacuum enclosure. Such flywheels can reach maximum speed ("charge") in a matter of minutes. The flywheel system is connected to a combination electric motor/generator.

A Flybrid Kinetic Energy Recovery System flywheel. Built for use on Formula 1 racing cars,
it is employed to recover and reuse kinetic energy captured during braking.

FES systems have relatively long lifetimes (lasting decades with little or no maintenance; full-cycle lifetimes quoted for flywheels range from in excess of $10^5$, up to $10^7$, cycles of use), high specific energy (100–130 W·h/kg, or 360–500 kJ/kg) and power density.

## Gravitational Potential Energy Storage with Solid Masses

Changing the altitude of solid masses can store or release energy via an elevating system driven by an electric motor/generator.

Companies such as Energy Cache and Advanced Rail Energy Storage (ARES) are working on this. ARES uses rails to move concrete weights up and down. Stratosolar proposes to use winches supported by buoyant platforms at an altitude of 20 kilometers, to raise and lower solid masses. Sink Float Solutions proposes to use winches supported by an ocean barge for taking advantage of a 4 km (13,000 ft) elevation difference between the surface and the seabed. ARES estimated a capital cost for the storage capacity of around 60% of pump storage hydroelectricity, Stratosolar $100/kWh and Sink Float Solutions $25/kWh (4000 m depth) and $50/kWh (with 2000 m depth).

Potential energy storage or gravity energy storage was under active development in 2013 in association with the California Independent System Operator. It examined the movement of earth-filled hopper rail cars driven by electric locomotives) from lower to higher elevations.

ARES claimed advantages including indefinite storage with no energy losses, low costs when earth/rocks are used and conservation of water resources.

## Thermal Storage

District heating accumulation tower from Theiss near Krems an der
Donau in Lower Austria with a thermal capacity of 2 GWh.

Thermal storage is the temporary storage or removal of heat. TES is practical because of water's large heat of fusion: the melting of one metric ton of ice (approximately one cubic metre in size) can capture 334 megajoules [MJ] (317,000 BTU) of thermal energy.

An example is Alberta, Canada's Drake Landing Solar Community, for which 97% of the year-round heat is provided by solar-thermal collectors on the garage roofs, with a borehole thermal energy store (BTES) being the enabling technology. STES projects often have paybacks in the four-to-six year range. In Braestrup, Denmark, the community's solar district heating system also utilizes STES, at a storage temperature of 65 °C (149 °F). A heat pump, which is run only when there is surplus wind power available on the national grid, is used to raise the temperature to 80 °C (176 °F) for distribution. When surplus wind generated electricity is not available, a gas-fired boiler is used. Twenty percent of Braestrup's heat is solar.

## Latent Heat Thermal Energy Storage (LHTES)

Latent heat thermal energy storage systems works with materials with high latent heat (heat of fusion) capacity, known as phase change materials (PCMs). The main advantage of these materials is that their latent heat storage capacity is much more than sensible heat. In a specific temperature range, phase changes from solid to liquid absorbs a large amount of thermal energy for later use.

## Electrochemical

## Rechargeable Battery

A rechargeable battery bank used as an uninterruptible power supply in a data center

A rechargeable battery, comprises one or more electrochemical cells. It is known as a 'secondary cell' because its electrochemical reactions are electrically reversible. Rechargeable batteries come in many different shapes and sizes, ranging from button cells to megawatt grid systems.

Rechargeable batteries have lower total cost of use and environmental impact than non-rechargeable (disposable) batteries. Some rechargeable battery types are available in the same form factors as disposables. Rechargeable batteries have higher initial cost but can be recharged very cheaply and used many times.

Common rechargeable battery chemistries include:

- Lead–acid battery: Lead acid batteries hold the largest market share of electric storage products. A single cell produces about 2V when charged. In the charged state the metallic lead negative electrode and the lead sulfate positive electrode are immersed in a dilute sulfuric acid ($H_2SO_4$) electrolyte. In the discharge process electrons are pushed out of the cell as lead sulfate is formed at the negative electrode while the electrolyte is reduced to water.

- Nickel–cadmium battery (NiCd): Uses nickel oxide hydroxide and metallic cadmium as electrodes. Cadmium is a toxic element, and was banned for most uses by the European Union in 2004. Nickel–cadmium batteries have been almost completely replaced by nickel–metal hydride (NiMH) batteries.

- Nickel–metal hydride battery (NiMH): First commercial types were available in 1989. These are now a common consumer and industrial type. The battery has a hydrogen-absorbing alloy for the negative electrode instead of cadmium.

- Lithium-ion battery: The choice in many consumer electronics and have one of the best energy-to-mass ratios and a very slow self-discharge when not in use.

- Lithium-ion polymer battery: These batteries are light in weight and can be made in any shape desired.

## Flow Battery

A flow battery operates by passing a solution over a membrane where ions are exchanged to charge/discharge the cell. Cell voltage is chemically determined by the Nernst equation and ranges, in practical applications, from 1.0 to 2.2 V. Its storage capacity is a function of the volume of the tanks holding the solution.

A flow battery is technically akin both to a fuel cell and an electrochemical accumulator cell. Commercial applications are for long half-cycle storage such as backup grid power.

## Supercapacitor

Supercapacitors, also called electric double-layer capacitors (EDLC) or ultracapacitors, are generic terms for a family of electrochemical capacitors that do not have conventional solid dielectrics. Capacitance is determined by two storage principles, double-layer capacitance and pseudocapacitance.

One of a fleet of electric capabuses powered by supercapacitors, at a quick-charge station-bus stop, in service during Expo 2010 Shanghai China. Charging rails can be seen suspended over the bus.

Supercapacitors bridge the gap between conventional capacitors and rechargeable batteries. They store the most energy per unit volume or mass (energy density) among capacitors. They support up to 10,000 farads/1.2 volt, up to 10,000 times that of electrolytic capacitors, but deliver or accept less than half as much power per unit time (power density).

While supercapacitors have specific energy and energy densities that are approximately 10% of batteries, their power density is generally 10 to 100 times greater. This results in much shorter charge/discharge cycles. Additionally, they will tolerate many more charge and discharge cycles than batteries.

Supercapacitors support a broad spectrum of applications, including:

- Low supply current for memory backup in static random-access memory (SRAM)

- Power for cars, buses, trains, cranes and elevators, including energy recovery from braking, short-term energy storage and burst-mode power delivery

## UltraBattery

The UltraBattery is a hybrid lead-acid cell and carbon-based ultracapacitor (or supercapacitor) invented by Australia's national research body, the Commonwealth Scientific and Industrial Research Organisation (CSIRO). The lead-acid cell and ultracapacitor share the sulfuric acid electrolyte and both are packaged into the same physical unit. The UltraBattery can be manufactured with similar physical and electrical characteristics to conventional lead-acid batteries making it possible to cost-effectively replace many lead-acid applications.

The UltraBattery tolerates high charge and discharge levels and endures large numbers of cycles, outperforming previous lead-acid cells by more than an order of magnitude. In hybrid-electric vehicle tests, millions of cycles have been achieved. The UltraBattery is also highly tolerant to the effects of sulfation compared with traditional lead-acid cells. This means it can operate continuously in partial state of charge whereas traditional lead-acid batteries are generally held at full charge between discharge events. It is generally electrically inefficient to fully charge a lead-acid battery so by decreasing time spent in the top region of charge the UltraBattery achieves high efficiencies, typically between 85 and 95% DC-DC.

The UltraBattery can work across a wide range of applications. The constant cycling and fast charging and discharging necessary for applications such as grid regulation and leveling and electric vehicles can damage chemical batteries, but are well handled by the ultracapacitive qualities of UltraBattery technology. The technology has been installed in Australia and the US on the megawatt scale, performing frequency regulation and renewable smoothing applications.

## Other Chemical

### Power to Gas

Power to gas is a technology which converts electricity into a gaseous fuel such as hydrogen or methane. The three commercial methods use electricity to reduce water into hydrogen and oxygen by means of electrolysis.

In the first method, hydrogen is injected into the natural gas grid or is used in transport or industry. The second method is to combine the hydrogen with carbon dioxide to produce methane using a methanation reaction such as the Sabatier reaction, or biological methanation, resulting in an extra energy conversion loss of 8%. The methane may then be fed into the natural gas grid. The third method uses the output gas of a wood gas generator or a biogas plant, after the biogas upgrader is mixed with the hydrogen from the electrolyzer, to upgrade the quality of the biogas.

### Hydrogen

The element hydrogen can be a form of stored energy. Hydrogen can produce electricity via a hydrogen fuel cell.

At penetrations below 20% of the grid demand, renewables do not severely change the economics; but beyond about 20% of the total demand, external storage becomes important. If these sources are used to make ionic hydrogen, they can be freely expanded. A 5-year community-based pilot program using wind turbines and hydrogen generators began in 2007 in the remote community of Ramea, Newfoundland and Labrador. A similar project began in 2004 on Utsira, a small Norwegian island.

Energy losses involved in the hydrogen storage cycle come from the electrolysis of water, liquification or compression of the hydrogen and conversion to electricity.

About 50 kWh (180 MJ) of solar energy is required to produce a kilogram of hydrogen, so the cost of the electricity is crucial. At $0.03/kWh, a common off-peak high-voltage line rate in the United States, hydrogen costs $1.50 a kilogram for the electricity, equivalent to $1.50/gallon for gasoline. Other costs include the electrolyzer plant, hydrogen compressors or liquefaction, storage and transportation.

Underground hydrogen storage is the practice of hydrogen storage in underground caverns, salt domes and depleted oil and gas fields. Large quantities of gaseous hydrogen have been stored in underground caverns by Imperial Chemical Industries for many years without any difficulties. The European Hyunder project indicated in 2013 that storage of wind and solar energy using underground hydrogen would require 85 caverns.

## Methane

Methane is the simplest hydrocarbon with the molecular formula $CH_4$. Methane is more easily stored and transported than hydrogen. Storage and combustion infrastructure (pipelines, gasometers, power plants) are mature.

Synthetic natural gas (syngas or SNG) can be created in a multi-step process, starting with hydrogen and oxygen. Hydrogen is then reacted with carbon dioxide in a Sabatier process, producing methane and water. Methane can be stored and later used to produce electricity. The resulting water is recycled, reducing the need for water. In the electrolysis stage oxygen is stored for methane combustion in a pure oxygen environment at an adjacent power plant, eliminating nitrogen oxides.

Methane combustion produces carbon dioxide ($CO_2$) and water. The carbon dioxide can be recycled to boost the Sabatier process and water can be recycled for further electrolysis. Methane production, storage and combustion recycles the reaction products.

The $CO_2$ has economic value as a component of an energy storage vector, not a cost as in carbon capture and storage.

## Power to Liquid

Power to liquid is similar to power to gas, however the hydrogen produced by electrolysis from wind and solar electricity isn't converted into gases such as methane but into liquids such as methanol. Methanol is easier in handling than gases and requires less safety precautions than hydrogen. It can be used for transportation, including aircraft, but also for industrial purposes or in the power sector.

## Biofuels

Various biofuels such as biodiesel, vegetable oil, alcohol fuels, or biomass can replace fossil fuels. Various chemical processes can convert the carbon and hydrogen in coal, natural gas, plant and animal biomass and organic wastes into short hydrocarbons suitable as replacements for existing hydrocarbon fuels. Examples are Fischer–Tropsch diesel, methanol, dimethyl ether and syngas. This diesel source was used extensively in World War II in Germany, which faced limited access to crude oil supplies. South Africa produces most of the country's diesel from coal for similar reasons. A long term oil price above US$35/bbl may make such large scale synthetic liquid fuels economical.

## Aluminium, Boron, Silicon, and Zinc

Aluminium, Boron, silicon, lithium, and zinc have been proposed as energy storage solutions.

## Other Chemical

The organic compound *norbornadiene* converts to *quadricyclane* upon exposure to light, storing solar energy as the energy of chemical bonds. A working system has been developed in Sweden as a molecular solar thermal system.

# Electrical Methods

## Capacitor

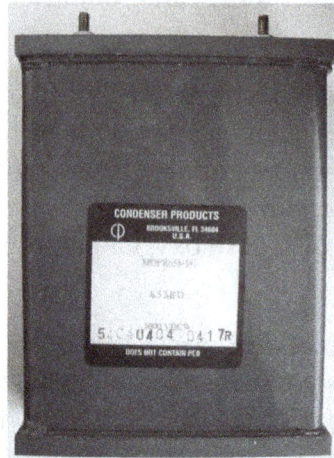

This mylar-film, oil-filled capacitor has very low inductance and low resistance, to provide the high-power (70 mega-watts) and the very high speed (1.2 microsecond) discharges needed to operate a dye laser.

A capacitor (originally known as a 'condenser') is a passive two-terminal electrical component used to store energy electrostatically. Practical capacitors vary widely, but all contain at least two electrical conductors (plates) separated by a dielectric (i.e., insulator). A capacitor can store electric energy when disconnected from its charging circuit, so it can be used like a temporary battery, or like other types of rechargeable energy storage system. Capacitors are commonly used in electronic devices to maintain power supply while batteries change. (This prevents loss of information in volatile memory.) Conventional capacitors provide less than 360 joules per kilogram, while a conventional alkaline battery has a density of 590 kJ/kg.

Capacitors store energy in an electrostatic field between their plates. Given a potential difference across the conductors (e.g., when a capacitor is attached across a battery), an electric field develops across the dielectric, causing positive charge (+Q) to collect on one plate and negative charge (-Q) to collect on the other plate. If a battery is attached to a capacitor for a sufficient amount of time, no current can flow through the capacitor. However, if an accelerating or alternating voltage is applied across the leads of the capacitor, a displacement current can flow.

Capacitance is greater given a narrower separation between conductors and when the conductors have a larger surface area. In practice, the dielectric between the plates emits a small amount of leakage current and has an electric field strength limit, known as the breakdown voltage. However, the effect of recovery of a dielectric after a high-voltage breakdown holds promise for a new generation of self-healing capacitors. The conductors and leads introduce undesired inductance and resistance.

Research is assessing the quantum effects of nanoscale capacitors for digital quantum batteries.

## Superconducting Magnetics

Superconducting magnetic energy storage (SMES) systems store energy in a magnetic field created by the flow of direct current in a superconducting coil that has been cooled to a temperature below its superconducting critical temperature. A typical SMES system includes a superconducting coil, power conditioning system and refrigerator. Once the superconducting coil is charged, the current does not decay and the magnetic energy can be stored indefinitely.

The stored energy can be released to the network by discharging the coil. The associated inverter/ rectifier accounts for about 2–3% energy loss in each direction. SMES loses the least amount of electricity in the energy storage process compared to other methods of storing energy. SMES systems offer round-trip efficiency greater than 95%.

Due to the energy requirements of refrigeration and the cost of superconducting wire, SMES is used for short duration storage such as improving power quality. It also has applications in grid balancing.

## Interseasonal Thermal Storage

Seasonal thermal energy storage (STES) allows heat or cold to be used months after it was collected from waste energy or natural sources. The material can be stored in contained aquifers, clusters of boreholes in geological substrates such as sand or crystalline bedrock, in lined pits filled with gravel and water, or water-filled mines.

## Applications

### Mills

The classic application before the industrial revolution was the control of waterways to drive water mills for processing grain or powering machinery. Complex systems of reservoirs and dams were constructed to store and release water (and the potential energy it contained) when required.

### Home Energy Storage

Home energy storage is expected to become increasingly common given the growing importance of distributed generation of renewable energies (especially photovoltaics) and the important share of energy consumption in buildings. To exceed a self-sufficiency of 40% in a household equipped with photovoltaics, energy storage is needed. Multiple manufacturers produce rechargeable battery systems for storing energy, generally to hold surplus energy from home solar/wind generation. Today, for home energy storage, Li-ion batteries are preferable to lead-acid ones given their similar cost but much better performance.

Tesla Motors produces two models of the Tesla Powerwall. One is a 10 kWh weekly cycle version for backup applications and the other is a 7 kWh version for daily cycle applications. In 2016, a limited version of the Telsa Powerpack 2 cost $398(US)/kWh to store electricity worth 12.5 cents/ kWh (US average grid price) making a positive return on investment doubtful unless electricity prices are higher than 30 cents/kWh.

Enphase Energy announced an integrated system that allows home users to store, monitor and manage electricity. The system stores 1.2 kWh hours of energy and 275W/500W power output.

Storing wind or solar energy using thermal energy storage though less flexible, is considerably less expensive than batteries. A simple 52-gallon electric water heater can store roughly 12 kWh of energy for supplementing hot water or space heating.

For purely financial purposes in areas where net metering is available, home generated electricity may be sold to the grid through a grid-tie inverter without the use of batteries for storage.

## Renewable Energy Storage

Construction of the Salt Tanks which provide efficient thermal energy storage so that output can be provided after the sun goes down, and output can be scheduled to meet demand requirements. The 280 MW Solana Generating Station is designed to provide six hours of energy storage. This allows the plant to generate about 38 percent of its rated capacity over the course of a year.

The 150 MW Andasol solar power station is a commercial parabolic trough solar thermal power plant, located in Spain. The Andasol plant uses tanks of molten salt to store solar energy so that it can continue generating electricity even when the sun isn't shining.

The largest source and the greatest store of renewable energy is provided by hydroelectric dams. A large reservoir behind a dam can store enough water to average the annual flow of a river between dry and wet seasons. A very large reservoir can store enough water to average the flow of a river between dry and wet years. While a hydroelectric dam does not directly store energy from intermittent sources, it does balance the grid by lowering its output and retaining its water when power is generated by solar or wind. If wind or solar generation exceeds the regions hydroelectric capacity, then some additional source of energy will be needed.

Many renewable energy sources (notably solar and wind) produce variable power. Storage systems can level out the imbalances between supply and demand that this causes. Electricity must be used as it is generated or converted immediately into storable forms.

The main method of electrical grid storage is pumped-storage hydroelectricity. Areas of the world such as Norway, Wales, Japan and the US have used elevated geographic features for reservoirs, using electrically powered pumps to fill them. When needed, the water passes through generators and converts the gravitational potential of the falling water into electricity. Pumped storage in Norway, which gets almost all its electricity from hydro, has an instantaneous capacity of 25–30 GW expandable to 60 GW—enough to be "Europe's battery".

Some forms of storage that produce electricity include pumped-storage hydroelectric dams, rechargeable batteries, thermal storage including molten salts which can efficiently store and release very large quantities of heat energy, and compressed air energy storage, flywheels, cryogenic systems and superconducting magnetic coils.

Surplus power can also be converted into methane (sabatier process) with stockage in the natural gas network.

In 2011, the Bonneville Power Administration in Northwestern United States created an experimental program to absorb excess wind and hydro power generated at night or during stormy periods that are accompanied by high winds. Under central control, home appliances absorb surplus energy by heating ceramic bricks in special space heaters to hundreds of degrees and by boosting the temperature of modified hot water heater tanks. After charging, the appliances provide home heating and hot water as needed. The experimental system was created as a result of a severe 2010 storm that overproduced renewable energy to the extent that all conventional power sources were shut down, or in the case of a nuclear power plant, reduced to its lowest possible operating level, leaving a large area running almost completely on renewable energy.

Another advanced method used at the former Solar Two project in the United States and the Solar Tres Power Tower in Spain uses molten salt to store thermal energy captured from the sun and then convert it and dispatch it as electrical power. The system pumps molten salt through a tower or other special conduits to be heated by the sun. Insulated tanks store the solution. Electricity is produced by turning water to steam that is fed to turbines.

Since the early 21st century batteries have been applied to utility scale load-leveling and frequency regulation capabilities.

In vehicle-to-grid storage, electric vehicles that are plugged into the energy grid can deliver stored electrical energy from their batteries into the grid when needed.

## Generation

Chemical fossil fuels (gas, oil, coal) remain the dominant form of energy storage for electricity generation, within natural gas becoming increasingly important.

## Air Conditioning

Thermal energy storage (TES) can be used for air conditioning. It is most widely used for cooling single large buildings and/or groups of smaller buildings. Commercial air conditioning systems are the biggest contributors to peak electrical loads. In 2009, thermal storage was used in over 3,300 buildings in over 35 countries. It works by creating ice at night and using the ice to for cooling during the hotter daytime periods.

The most popular technique is ice storage, which requires less space than water and is less costly than fuel cells or flywheels. In this application, a standard chiller runs at night to produce an ice pile. Water then circulates through the pile during the day to chill water that would normally be the chiller's daytime output.

A partial storage system minimizes capital investment by running the chillers nearly 24 hours a day. At night, they produce ice for storage and during the day they chill water. Water circulating through the melting ice augments the production of chilled water. Such a system makes ice for 16 to 18 hours a day and melts ice for six hours a day. Capital expenditures are reduced because the chillers can be just 40 - 50% of the size needed for a conventional, no-storage design. Storage sufficient to store half a day's available heat is usually adequate.

A full storage system shuts off the chillers during peak load hours. Capital costs are higher, as such a system requires larger chillers and a larger ice storage system.

This ice is produced when electrical utility rates are lower. Off-peak cooling systems can lower energy costs. The U.S. Green Building Council has developed the Leadership in Energy and Environmental Design (LEED) program to encourage the design of reduced-environmental impact buildings. Off-peak cooling may help toward LEED Certification.

Thermal storage for heating is less common than for cooling. An example of thermal storage is storing solar heat to be used for heating at night.

Latent heat can also be stored in technical phase change materials (PCMs). These can be encapsulated in wall and ceiling panels, to moderate room temperatures.

## Transport

Liquid hydrocarbon fuels are the most commonly used forms of energy storage for use in transportation. Other energy carriers such as hydrogen can be used to avoid producing greenhouse gases.

## Electronics

Capacitors are widely used in electronic circuits for blocking direct current while allowing alternating current to pass. In analog filter networks, they smooth the output of power supplies. In resonant circuits they tune radios to particular frequencies. In electric power transmission systems they stabilize voltage and power flow.

## Use Cases

The United States Department of Energy International Energy Storage Database (IESDB), is a free-access database of energy storage projects and policies funded by the United States Department of Energy Office of Electricity and Sandia National Labs.

## Storage Capacity

Storage capacity is the amount of energy extracted from a power plant energy storage system; usually measured in joules or kilowatt-hours and their multiples, it may be given in number of hours of electricity production at power plant nameplate capacity; when storage is of primary type (i.e., thermal or pumped-water), output is sourced only with the power plant embedded storage system.

## Economics

The economics of Energy Storage strictly depends on the reserve service requested, and several uncertainty factors affect the profitability of Energy Storage. Therefore, not every Energy Storage is technically and economically suitable for the storage of several MWh, and the optimal size of the Energy Storage is market and location dependent.

Moreover, ESS are affected by several risks, e.g.:

1) Techno-economic risks, which are related to the specific technology;

2) Market risks, which are the factors that affect the electricity supply system;

3) Regulation and policy risks.

Therefore, traditional techniques based on deterministic Discounted Cash Flow (DCF) for the investment appraisal are not fully adequate to evaluate these risks and uncertainties and the investor's flexibility to deal with them. Hence, the literature recommends to assess the value of risks and uncertainties through the Real Option Analysis (ROA), which is a valuable method in uncertain contexts.

The economic valuation of large-scale applications (including pumped hydro storage and compressed air) considers benefits including: wind curtailment avoidance, grid congestion avoidance, price arbitrage and carbon free energy delivery. In one technical assessment by the Carnegie Mellon Electricity Industry Centre, economic goals could be met with batteries if energy storage were achievable at a capital cost of $30 to $50 per kilowatt-hour of storage capacity.

A metric for calculating the energy efficiency of storage systems is Energy Storage On Energy Invested (ESOI) which is the useful energy used to make the storage system divided into the lifetime energy storage. For lithium ion batteries this is around 10, and for lead acid batteries it is about 2. Other forms of storage such as pumped hydroelectric storage generally have higher ESOI, such as 210.

## Research

### Germany

In 2013, the German Federal government has allocated €200M (approximately US$270M) for advanced research, as well as providing a further €50M to subsidize battery storage for use with residential rooftop solar panels, according to a representative of the German Energy Storage Association.

Siemens AG commissioned a production-research plant to open in 2015 at the *Zentrum für Sonnenenergie und Wasserstoff (ZSW,* the German Center for Solar Energy and Hydrogen Research in the State of Baden-Württemberg), a university/industry collaboration in Stuttgart, Ulm and Widderstall, staffed by approximately 350 scientists, researchers, engineers, and technicians. The plant develops new near-production manufacturing materials and processes (NPMM&P) using a computerized Supervisory Control and Data Acquisition (SCADA) system. Its goals will enable the expansion of rechargeable battery production with both increased quality and reduced manufacturing costs.

## United States

In 2014, research and test centers opened to evaluate energy storage technologies. Among them was the Advanced Systems Test Laboratory at the University of Wisconsin at Madison in Wisconsin State, which partnered with battery manufacturer Johnson Controls. The laboratory was created as part of the university's newly opened Wisconsin Energy Institute. Their goals include the evaluation of state-of-the-art and next generation electric vehicle batteries, including their use as grid supplements.

The State of New York unveiled its New York Battery and Energy Storage Technology (NY-BEST) Test and Commercialization Center at Eastman Business Park in Rochester, New York, at a cost of $23 million for its almost 1,700 m$^2$ laboratory. The center includes the Center for Future Energy Systems, a collaboration between Cornell University of Ithaca, New York and the Rensselaer Polytechnic Institute in Troy, New York. NY-BEST tests, validates and independently certifies diverse forms of energy storage intended for commercial use.

## United Kingdom

In the United Kingdom, some fourteen industry and government agencies allied with seven British universities in May 2014 to create the SUPERGEN Energy Storage Hub in order to assist in the coordination of energy storage technology research and development.

## Superconducting Magnetic Energy Storage

Superconducting Magnetic Energy Storage (SMES) systems store energy in the magnetic field created by the flow of direct current in a superconducting coil which has been cryogenically cooled to a temperature below its superconducting critical temperature.

A typical SMES system includes three parts: superconducting coil, power conditioning system and cryogenically cooled refrigerator. Once the superconducting coil is charged, the current will not decay and the magnetic energy can be stored indefinitely.

The stored energy can be released back to the network by discharging the coil. The power conditioning system uses an inverter/rectifier to transform alternating current (AC) power to direct current or convert DC back to AC power. The inverter/rectifier accounts for about 2–3% energy loss in each direction. SMES loses the least amount of electricity in the energy storage process compared to other methods of storing energy. SMES systems are highly efficient; the round-trip efficiency is greater than 95%.

Due to the energy requirements of refrigeration and the high cost of superconducting wire, SMES is currently used for short duration energy storage. Therefore, SMES is most commonly devoted to improving power quality.

## Advantages Over other Energy Storage Methods

There are several reasons for using superconducting magnetic energy storage instead of other energy storage methods. The most important advantage of SMES is that the time delay during charge and discharge is quite short. Power is available almost instantaneously and very high power output can be provided for a brief period of time. Other energy storage methods, such as pumped hydro or compressed air, have a substantial time delay associated with the energy conversion of stored mechanical energy back into electricity. Thus if demand is immediate, SMES is a viable option. Another advantage is that the loss of power is less than other storage methods because electric currents encounter almost no resistance. Additionally the main parts in a SMES are motionless, which results in high reliability.

## Current Use

There are several small SMES units available for commercial use and several larger test bed projects. Several 1 MW·h units are used for power quality control in installations around the world, especially to provide power quality at manufacturing plants requiring ultra-clean power, such as microchip fabrication facilities.

These facilities have also been used to provide grid stability in distribution systems. SMES is also used in utility applications. In northern Wisconsin, a string of distributed SMES units were deployed to enhance stability of a transmission loop. The transmission line is subject to large, sudden load changes due to the operation of a paper mill, with the potential for uncontrolled fluctuations and voltage collapse.

The Engineering Test Model is a large SMES with a capacity of approximately 20 MW·h, capable of providing 40 MW of power for 30 minutes or 10 MW of power for 2 hours.

## Calculation of Stored Energy

The magnetic energy stored by a coil carrying a current is given by one half of the inductance of the coil times the square of the current.

$$E = \frac{1}{2}LI^2$$

Where

      $E$ = energy measured in joules

      $L$ = inductance measured in henries

      $I$ = current measured in amperes

Now let's consider a cylindrical coil with conductors of a rectangular cross section. The mean radius of coil is $R$. $a$ and $b$ are width and depth of the conductor. $f$ is called form function which is different for different shapes of coil. $\xi$ (xi) and $\delta$ (delta) are two parameters to characterize the dimensions of the coil. We can therefore write the magnetic energy stored in such a cylindrical coil as shown. This energy is a function of coil dimensions, number of turns and carrying current.

$$E = \frac{1}{2} R N^2 I^2 f\left(\xi, \delta\right)$$

Where

$E$ = energy measured in joules

$I$ = current measured in amperes

$f(\xi,\delta)$ = form function, joules per ampere-meter

$N$ = number of turns of coil

## Solenoid Versus Toroid

Besides the properties of the wire, the configuration of the coil itself is an important issue from a mechanical engineering aspect. There are three factors which affect the design and the shape of the coil - they are: Inferior strain tolerance, thermal contraction upon cooling and Lorentz forces in a charged coil. Among them, the strain tolerance is crucial not because of any electrical effect, but because it determines how much structural material is needed to keep the SMES from breaking. For small SMES systems, the optimistic value of 0.3% strain tolerance is selected. Toroidal geometry can help to lessen the external magnetic forces and therefore reduces the size of mechanical support needed. Also, due to the low external magnetic field, toroidal SMES can be located near a utility or customer load.

For small SMES, solenoids are usually used because they are easy to coil and no pre-compression is needed. In toroidal SMES, the coil is always under compression by the outer hoops and two disks, one of which is on the top and the other is on the bottom to avoid breakage. Currently, there is little need for toroidal geometry for small SMES, but as the size increases, mechanical forces become more important and the toroidal coil is needed.

The older large SMES concepts usually featured a low aspect ratio solenoid approximately 100 m in diameter buried in earth. At the low extreme of size is the concept of micro-SMES solenoids, for energy storage range near 1 MJ.

## Low-temperature versus High-temperature Superconductors

Under steady state conditions and in the superconducting state, the coil resistance is negligible. However, the refrigerator necessary to keep the superconductor cool requires electric power and this refrigeration energy must be considered when evaluating the efficiency of SMES as an energy storage device.

Although the high-temperature superconductor (HTSC) has higher critical temperature, flux lattice melting takes place in moderate magnetic fields around a temperature lower than this critical temperature. The heat loads that must be removed by the cooling system include conduction through the support system, radiation from warmer to colder surfaces, AC losses in the conductor (during charge and discharge), and losses from the cold–to-warm power leads that connect the cold coil to the power conditioning system. Conduction and radiation losses are minimized by proper design of thermal surfaces. Lead losses can be minimized by good design of the leads. AC losses depend on the design of the conductor, the duty cycle of the device and the power rating.

The refrigeration requirements for HTSC and low-temperature superconductor (LTSC) toroidal coils for the baseline temperatures of 77 K, 20 K, and 4.2 K, increases in that order. The refrigeration requirements here is defined as electrical power to operate the refrigeration system. As the stored energy increases by a factor of 100, refrigeration cost only goes up by a factor of 20. Also, the savings in refrigeration for an HTSC system is larger (by 60% to 70%) than for an LTSC systems.

## Cost

Whether HTSC or LTSC systems are more economical depends because there are other major components determining the cost of SMES: Conductor consisting of superconductor and copper stabilizer and cold support are major costs in themselves. They must be judged with the overall efficiency and cost of the device. Other components, such as vacuum vessel insulation, has been shown to be a small part compared to the large coil cost. The combined costs of conductors, structure and refrigerator for toroidal coils are dominated by the cost of the superconductor. The same trend is true for solenoid coils. HTSC coils cost more than LTSC coils by a factor of 2 to 4. We expect to see a cheaper cost for HTSC due to lower refrigeration requirements but this is not the case. So, why is the HTSC system more expensive?

To gain some insight consider a breakdown by major components of both HTSC and LTSC coils corresponding to three typical stored energy levels, 2, 20 and 200 MW·h. The conductor cost dominates the three costs for all HTSC cases and is particularly important at small sizes. The principal reason lies in the comparative current density of LTSC and HTSC materials. The critical current of HTSC wire is lower than LTSC wire generally in the operating magnetic field, about 5 to 10 teslas (T). Assume the wire costs are the same by weight. Because HTSC wire has lower $(J_c)$ value than LTSC wire, it will take much more wire to create the same inductance. Therefore, the cost of wire is much higher than LTSC wire. Also, as the SMES size goes up from 2 to 20 to 200 MW·h, the LTSC conductor cost also goes up about a factor of 10 at each step. The HTSC conductor cost rises a little slower but is still by far the costliest item.

The structure costs of either HTSC or LTSC go up uniformly (a factor of 10) with each step from 2 to 20 to 200 MW·h. But HTSC structure cost is higher because the strain tolerance of the HTSC (ceramics cannot carry much tensile load) is less than LTSC, such as $Nb_3Ti$ or $Nb_3Sn$, which demands more structure materials. Thus, in the very large cases, the HTSC cost can not be offset by simply reducing the coil size at a higher magnetic field.

It is worth noting here that the refrigerator cost in all cases is so small that there is very little percentage savings associated with reduced refrigeration demands at high temperature. This means that if a HTSC, BSCCO for instance, works better at a low temperature, say 20K, it will certainly be operated there. For very small SMES, the reduced refrigerator cost will have a more significant positive impact.

Clearly, the volume of superconducting coils increases with the stored energy. Also, we can see that the LTSC torus maximum diameter is always smaller for a HTSC magnet than LTSC due to higher magnetic field operation. In the case of solenoid coils, the height or length is also smaller for HTSC coils, but still much higher than in a toroidal geometry (due to low external magnetic field).

An increase in peak magnetic field yields a reduction in both volume (higher energy density) and cost (reduced conductor length). Smaller volume means higher energy density and cost is reduced due to the decrease of the conductor length. There is an optimum value of the peak magnetic field, about 7 T in this case. If the field is increased past the optimum, further volume reductions are possible with minimal increase in cost. The limit to which the field can be increased is usually not economic but physical and it relates to the impossibility of bringing the inner legs of the toroid any closer together and still leave room for the bucking cylinder.

The superconductor material is a key issue for SMES. Superconductor development efforts focus on increasing Jc and strain range and on reducing the wire manufacturing cost.

## Technical Challenges

The energy content of current SMES systems is usually quite small. Methods to increase the energy stored in SMES often resort to large-scale storage units. As with other superconducting applications, cryogenics are a necessity. A robust mechanical structure is usually required to contain the very large Lorentz forces generated by and on the magnet coils. The dominant cost for SMES is the superconductor, followed by the cooling system and the rest of the mechanical structure.

### Mechanical Support

Needed because of Lorentz forces.

### Size

To achieve commercially useful levels of storage, around 1 GW·h (3.6 TJ), a SMES installation would need a loop of around 100 miles (160 km). This is traditionally pictured as a circle, though in practice it could be more like a rounded rectangle. In either case it would require access to a significant amount of land to house the installation.

### Manufacturing

There are two manufacturing issues around SMES. The first is the fabrication of bulk cable suitable to carry the current. Most of the superconducting materials found to date are relatively delicate ceramics, making it difficult to use established techniques to draw extended lengths of superconducting wire. Much research has focussed on layer deposit techniques, applying a thin film of material onto a stable substrate, but this is currently only suitable for small-scale electrical circuits.

### Infrastructure

The second problem is the infrastructure required for an installation. Until room-temperature superconductors are found, the 100 mile (160 km) loop of wire would have to be contained within a vacuum flask of liquid nitrogen. This in turn would require stable support, most commonly envisioned by burying the installation.

## Critical Magnetic Field

Above a certain field strength, known as the critical field, the superconducting state is destroyed.

## Critical Current

In general power systems look to maximize the current they are able to handle. This makes any losses due to inefficiencies in the system relatively insignificant. Unfortunately, large currents may generate magnetic fields greater than the critical field due to Ampere's Law. Current materials struggle, therefore, to carry sufficient current to make a commercial storage facility economically viable.

Several issues at the onset of the technology have hindered its proliferation:

1. Expensive refrigeration units and high power cost to maintain operating temperatures

2. Existence and continued development of adequate technologies using normal conductors

These still pose problems for superconducting applications but are improving over time. Advances have been made in the performance of superconducting materials. Furthermore, the reliability and efficiency of refrigeration systems has improved significantly.

## Natural Gas Storage

Natural gas, like many other commodities, can be stored for an indefinite period of time in natural gas storage facilities for later consumption.

## Usage

Gas storage is principally used to meet load variations. Gas is injected into storage during periods of low demand and withdrawn from storage during periods of peak demand. It is also used for a variety of secondary purposes, including:

- Balancing the flow in pipeline systems. This is performed by mainline transmission pipeline companies to maintain operational integrity of the pipelines, by ensuring that the pipeline pressures are kept within design parameters.

- Maintaining contractual balance. Shippers use stored gas to maintain the volume they deliver to the pipeline system and the volume they withdraw. Without access to such storage facilities, any imbalance situation would result in a hefty penalty.

- Leveling production over periods of fluctuating demand. Producers use storage to store any gas that is not immediately marketable, typically over the summer when demand is low and deliver it in the winter months when the demand is high.

- Market speculation. Producers and marketers use gas storage as a speculative tool, storing gas when they believe that prices will increase in the future and then selling it when it does reach those levels.

- Insuring against any unforeseen accidents. Gas storage can be used as an insurance that may affect either production or delivery of natural gas. These may include natural factors such as hurricanes, or malfunction of production or distribution systems.

- Meeting regulatory obligations. Gas storage ensures to some extent the reliability of gas supply to the consumer at the lowest cost, as required by the regulatory body. This is why the regulatory body monitors storage inventory levels.

- Reducing price volatility. Gas storage ensures commodity liquidity at the market centers. This helps contain natural gas price volatility and uncertainty.

- Offsetting changes in natural gas demands. Gas storage facilities are gaining more importance due to changes in natural gas demands. First, traditional supplies that once met the winter peak demand are now unable to keep pace. Second, there is a growing summer peak demand on natural gas, due to electric generation via gas fired power plants.

**Underground Natural Gas Storage Working Gas Capacity in North America**

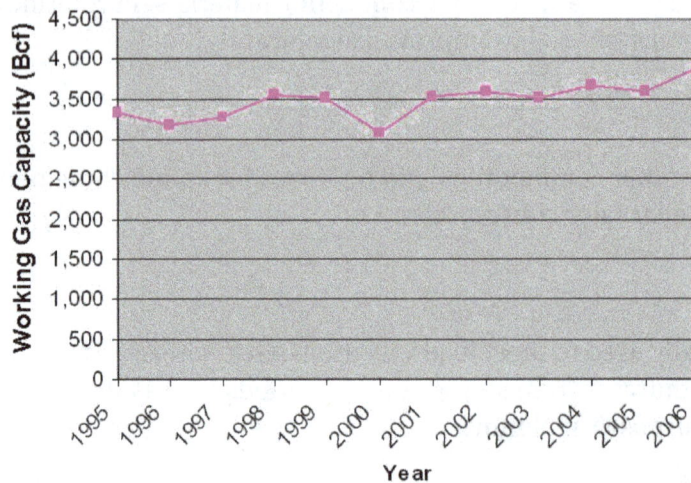

## Measures and Definitions

Characteristics of underground storage facilities need to be defined and measured. A number of volumetric measures have been put in place for that purpose:

- Total gas storage capacity: It is the maximum volume of natural gas that can be stored at the storage facility. It is determined by several physical factors such as the reservoir volume, and also on the operating procedures and engineering methods used.

- Total gas in storage: It is the total volume of gas in storage at the facility at a particular time.

- Base gas (also referred to as cushion gas): It is the volume of gas that is intended as permanent inventory in a storage reservoir to maintain adequate pressure and deliverability rates throughout the withdrawal season.

- Working gas capacity: It is the total gas storage capacity minus the base gas.

- Working gas: It is the total gas in storage minus the base gas. Working gas is the volume of gas available to the market place at a particular time.

- Physically unrecoverable gas: The amount of gas that becomes permanently embedded in the formation of the storage facility and that can never be extracted.

- Cycling rate: It is the average number of times a reservoir's working gas volume can be turned over during a specific period of time. Typically the period of time used is one year.

- Deliverability: It is a measure of the amount of gas that can be delivered (withdrawn) from a storage facility on a daily basis. It is also referred to as the deliverability rate, withdrawal rate, or withdrawal capacity and is usually expressed in terms of millions of cubic feet of gas per day (MMcf/day) that can be delivered.

- Injection capacity (or rate): It is the amount of gas that can be injected into a storage facility on a daily basis. It can be thought of as the complement of the deliverability. Injection rate is also typically measured in millions of cubic feet of gas that can be delivered per day (MMcf/day).

The measurements above are not fixed for a given storage facility. For example, deliverability depends on several factors including the amount of gas in the reservoir and the pressure etc. Generally, a storage facility's deliverability rate varies directly with the total amount of gas in the reservoir. It is at its highest when the reservoir is full and declines as gas is withdrawn. The injection capacity of a storage facility is also variable and depends on factors similar to those that affect deliverability. The injection rate varies inversely with the total amount of gas in storage. It is at its highest when the reservoir is nearly empty and declines as more gas is injected. The storage facility operator may also change operational parameters. This would allow, for example, the storage capacity maximum to be increased, the withdrawal of base gas during very high demand or reclassifying base gas to working gas if technological advances or engineering procedures allow.

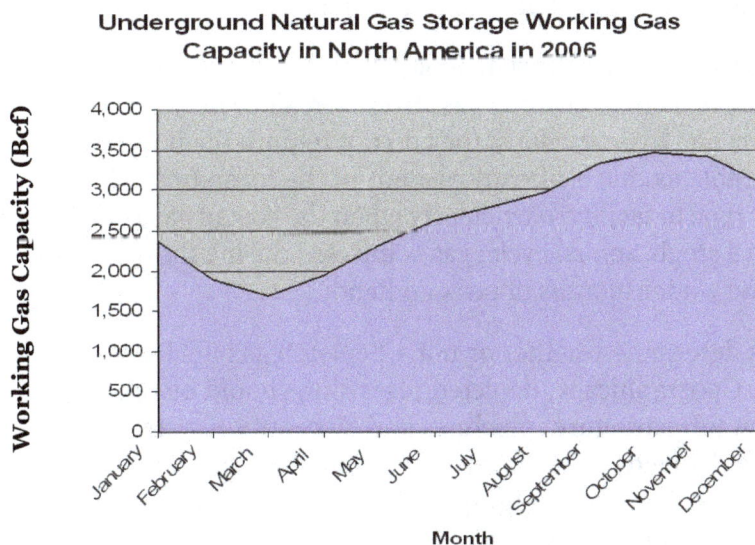

Underground Natural Gas Storage Working Gas Capacity in North America in 2006

## Types

The most important type of gas storage is in underground reservoirs. There are three principal types — depleted gas reservoirs, aquifer reservoirs and salt cavern reservoirs. Each of these types has distinct physical and economic characteristics which govern the suitability of a particular type of storage type for a given application.

Equipment of an underground natural gas storage facility in the Czech Republic near the town of Milín.

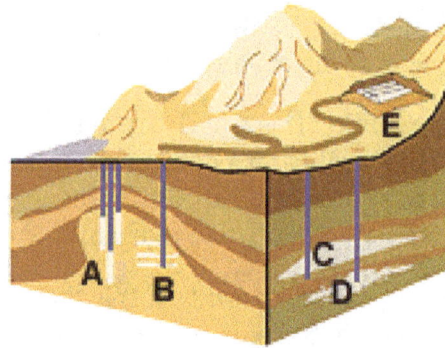

Natural gas is stored in underground (A) salt formations, (C) aquifer reservoirs and (D) depleted reservoirs.

## Depleted Gas Reservoir

These are the most prominent and common form of underground storage. They are the reservoir formations of natural gas fields that have produced all their economically recoverable gas. The depleted reservoir formation is readily capable of holding injected natural gas. Using such a facility is economically attractive because it allows the re-use, with suitable modification, of the extraction and distribution infrastructure remaining from the productive life of the gas field which reduces the start-up costs. Depleted reservoirs are also attractive because their geological and physical characteristics have already been studied by geologists and petroleum engineers and are usually well known. Consequently, depleted reservoirs are generally the cheapest and easiest to develop, operate, and maintain of the three types of underground storage.

In order to maintain working pressures in depleted reservoirs, about 50 percent of the natural gas in the formation must be kept as cushion gas. However, since depleted reservoirs were previously filled with natural gas and hydrocarbons, they do not require the injection of gas that will become physically unrecoverable as this is already present in the formation. This provides a further economic boost for this type of facility, particularly when the cost of gas is high. Typically, these facilities are operated on a single annual cycle; gas is injected during the off-peak summer months and withdrawn during the winter months of peak demand.

A number of factors determine whether or not a depleted gas field will make an economically viable storage facility. Geographically, depleted reservoirs should be relatively close to gas markets and to transportation infrastructure (pipelines and distribution systems) which will connect them to that market. Since the fields were at one time productive and connected to infrastructure, distance from market is the dominant geographical factor. Geologically, it is preferred that depleted reservoir formations have high porosity and permeability. The porosity of the formation is one of the factors that determines the amount of natural gas the reservoir is able to hold. Permeability is a measure of the rate at which natural gas flows through the formation and ultimately determines the rate of injection and withdrawal of gas from storage.

**Working Gas Capacity by Storage Facility Type**

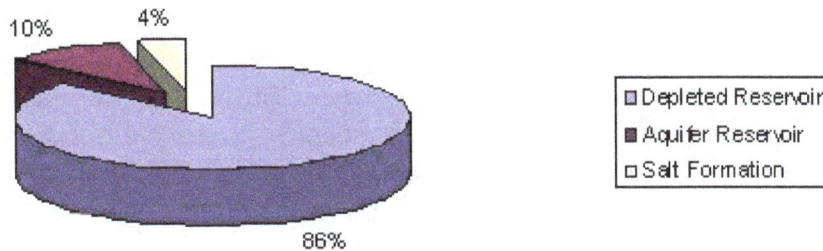

## Aquifer Reservoir

Aquifers are underground, porous and permeable rock formations that act as natural water reservoirs. In some cases they can be used for natural gas storage. Usually these facilities are operated on a single annual cycle as with depleted reservoirs. The geological and physical characteristics of aquifer formation are not known ahead of time and a significant investment has to go into investigating these and evaluating the aquifer's suitability for natural gas storage.

If the aquifer is suitable, all of the associated infrastructure must be developed from scratch, increasing the development costs compared to depleted reservoirs. This includes installation of wells, extraction equipment, pipelines, dehydration facilities, and possibly compression equipment. Since the aquifer initially contains water there is little or no naturally occurring gas in the formation and of the gas injected some will be physically unrecoverable. As a result, aquifer storage typically requires significantly more cushion gas than depleted reservoirs; up to 80% of the total gas volume. Most aquifer storage facilities were developed when the price of natural gas was low, meaning this cushion gas was inexpensive to sacrifice. With rising gas prices aquifer storage becomes more expensive to develop.

A consequence of the above factors is that developing an aquifer storage facility is usually time consuming and expensive. Aquifers are generally the least desirable and most expensive type of natural gas storage facility.

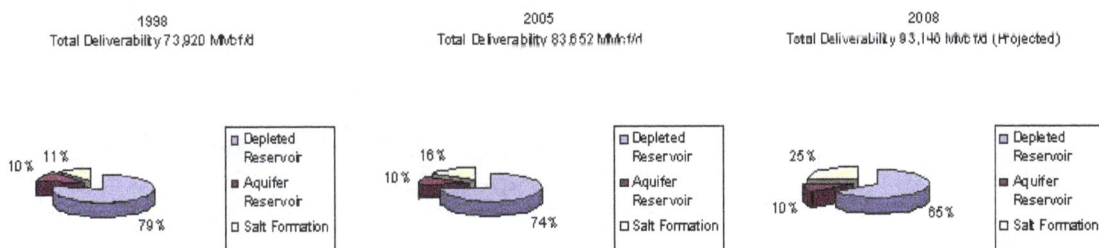

Total Deliverability from Natural Gas Storage by Type of Facility, 1998, 2005, 2008

## Salt Formation

Underground salt formations are well suited to natural gas storage. Salt caverns allow very little of the injected natural gas to escape from storage unless specifically extracted. The walls of a salt cavern are strong and impervious to gas over the lifespan of the storage facility.

Once a salt feature is discovered and found to be suitable for the development of a gas storage facility a cavern is created within the salt feature. This is done by the process of solution mining. Fresh water is pumped down a borehole into the salt. Some of the salt is dissolved leaving a void and the water, now saline, is pumped back to the surface. The process continues until the cavern is the desired size. Once created, a salt cavern offers an underground natural gas storage vessel with very high deliverability. Cushion gas requirements are low, typically about 33 percent of total gas capacity.

Salt caverns are usually much smaller than depleted gas reservoir and aquifer storage facilities. A salt cavern facility may occupy only one one-hundredth of the area taken up by a depleted gas reservoir facility. Consequently, salt caverns cannot hold the large volumes of gas necessary to meet base load storage requirements. Deliverability from salt caverns is, however, much higher than for either aquifers or depleted reservoirs. This allows the gas stored in a salt cavern to be withdrawn and replenished more readily and quickly. This quick cycle-time is useful in emergency situations or during short periods of unexpected demand surges.

Although construction is more costly than depleted field conversions when measured on the basis of dollars per thousand cubic feet of working gas, the ability to perform several withdrawal and injection cycles each year reduces the effective cost.

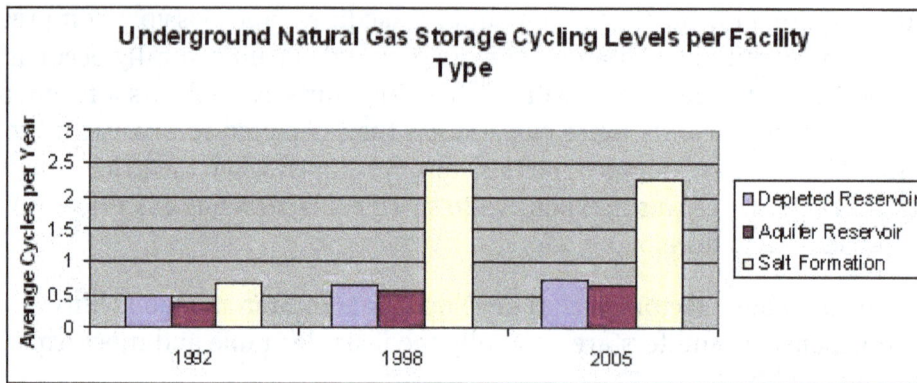

Datasource.

| Gas Storage Facility Operations | | | |
|---|---|---|---|
| Type | Cushion Gas | Injection Period (Days) | Withdrawal Period (Days) |
| Depleted Reservoir | 50% | 200-250 | 100-150 |
| Aquifer Reservoir | 50%-80% | 200-250 | 100-150 |
| Salt Formation | 20%-30% | 20-40 | 10-20 |

## Other

There are also other types of storage such as:

## LNG

A liquefied natural gas storage tank in Massachusetts

LNG facilities provide delivery capacity during peak periods when market demand exceeds pipe-line deliverability. LNG storage tanks possess a number of advantages over underground storage. As a liquid at approximately −163 °C (−260 °F), it occupies about 600 times less space than gas stored underground, and it provides high deliverability at very short notice because LNG storage facilities are generally located close to market and can be trucked to some customers avoiding pipeline tolls. There is no requirement for cushion gas and it allows access to a global supply. LNG facilities are, however, more expensive to build and maintain than developing new underground storage facilities.

## Pipeline Capacity

Gas can be temporarily stored in the pipeline system itself, through a process called line packing. This is done by packing more gas into the pipeline by an increase in the pressure. During periods of high demand, greater quantities of gas can be withdrawn from the pipeline in the market area, than is injected at the production area. The process of line packing is usually performed during off peak times to meet the next day's peaking demands. This method, however, only provides a temporary short-term substitute for traditional underground storage.

## Gasholders

Gas can be stored above ground in a gasholder (or gasometer), largely for balancing, not long-term storage, and this has been done since Victorian times. These store gas at district pressure, meaning that they can provide extra gas very quickly at peak times. Gasholders are perhaps most used in the United Kingdom and Germany. There are two kinds of gasholder — column-guided, which are guided up by a large frame that is always visible, regardless of the position of the holder; and spiral-guided, which have no frame and are guided up by concentric runners in the previous lift.

An older column-guided gasholder in West Ham, London

1960s-built spiral-guided gasholders in Hunslet, Leeds

Perhaps the most famous British gasholder is the large column-guided "Oval gasholders" that overlooks The Oval cricket ground in London. Gasholders were built in the United Kingdom from early Victorian times; many, such as Kings Cross in London and St. Marks Street in Kingston upon Hull are so old that they are entirely riveted, as their construction predates the use of welding in construction. The last to be built in the UK was in 1983.

## Owners

### Interstate Pipeline Companies

Interstate pipeline companies rely heavily on underground storage to perform load balancing and system supply management on their long-haul transmission lines. FERC regulations though demand that these companies open up the remainder of their capacity not used for that purpose to third parties. Twenty-five interstate companies currently operate 172 underground natural gas storage facilities. In 2005, their facilities accounted for about 43 percent of overall storage deliverability and 55 percent of working gas capacity in the US. These operators include the Columbia Gas Transmission Company, Dominion Gas Transmission Company, The National Fuel Gas Supply Company, Natural Gas Pipeline of America, Texas Gas Transmission Company, Southern Star Central Pipeline Company, TransCanada Corporation.

## Intrastate Pipeline Companies and Local Distribution Companies

Intrastate pipeline companies use storage facilities for operational balancing and system supply as well as to meet the energy demand of end-use customers. LDCs generally use gas from storage to serve customers directly. This group operates 148 underground storage sites and account for 40 percent of overall storage deliverability and 32 percent of working gas capacity in the US. These operators include Consumers Energy Company and the Northern Illinois Gas Company (Nicor), in the US and Enbridge and Union Gas in Canada.

## Independent Storage Service Providers

The deregulation activity in the underground gas storage arena has attracted independent storage service providers to develop storage facilities. The capacity made available would then be leased to third-party customers such as marketers and electricity generators. It is expected that in the future, this group would take more market share, as more deregulation takes place. Currently in the US, this group accounts for 18 percent of overall storage deliverability and 13 percent of working gas capacity in the US.

| Underground Natural Gas Storage by Type of Owner, 2005 | | | |
|---|---|---|---|
| **Type of Owner** | **Number of Sites** | **Working Gas Capacity (Bcf)** | **Daily Deliverability (MMcf)** |
| Interstate Pipeline | 172 | 2,197 | 35,830 |
| Intrastate & LDC | 148 | 1,292 | 33,121 |
| Independent | 74 | 521 | 14,681 |

## Location and Distribution

### Europe

As of January 2011, there are 124 underground storage facilities in Europe.

### United States

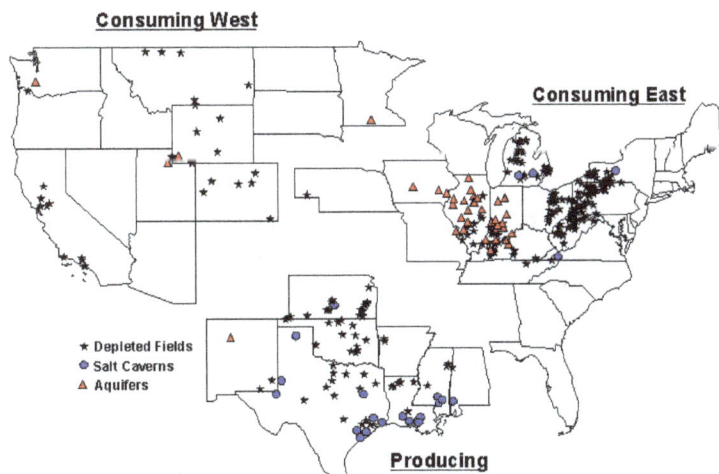

The United States is typically broken out into three main regions when it comes to gas consumption and production. These are the consuming East, the consuming West and the producing South.

## Consuming East

The consuming east region, particularly the states in the northern part, heavily rely on stored gas to meet the peak demand during the cold winter months. Due to the prevailing cold winters, large population centers and developed infrastructure, it is not surprising that this region has the highest level of working gas storage capacity of the other regions and the largest number of storage sites, mainly in depleted reservoirs. In addition to underground storage, LNG is increasingly playing a crucial role in providing supplemental backup and/or peaking supply to LDCs on a short term basis. Although the total capacity for these LNG facilities does not match those of underground storage in scale, the short term high deliverability makes up for that.

## Consuming West

The consuming west region has the smallest share of gas storage both in terms of the number of sites as well as gas capacity/deliverability. Storage in this area is mostly used to allow domestic and Albertan gas, coming from Canada, to flow at a rather constant rate.

## Producing South

The producing south's storage facilities are linked to the market centers and play a crucial role in the efficient export, transmission and distribution of natural gas produced to the consuming regions. These storage facilities allow the storage of gas that is not immediately marketable to be stored for later use.

| Underground Natural Gas Storage by Region, 2000 | | | |
|---|---|---|---|
| Region | Number of Sites | Working Gas Capacity (Bcf) | Daily Deliverability (MMcf) |
| East | 280 | 2,045 | 39,643 |
| West | 37 | 628 | 9,795 |
| South | 98 | 1,226 | 28,296 |

## Canada

In Canada, the maximum working gas stored was $456 \times 10^9$ cu ft ($1.29 \times 10^{10}$ m$^3$) in 2006. Alberta storage accounts for 47.5 percent of the total working gas volume. It is followed by Ontario which accounts for 39.1 percent, British Columbia which accounts for 7.6 percent, Saskatchewan which accounts for 5.1 percent and finally Quebec which accounts for 0.9 percent.

## Regulation and Deregulation

### United States

Interstate pipeline companies in the US are subject to the jurisdiction of the Federal Energy Regulatory Commission (FERC). Prior to 1992, these companies owned all the gas that flowed through their systems. This also included gas in their storage facility, over which they had complete control. Then FERC Order 636 was implemented. This required the companies to operate their facilities, including gas storage on an open access basis. For gas storage, this meant that these companies could only reserve the capacity needed to maintain system integrity. The rest of the capacity would be available for leasing to third parties on a nondiscriminatory basis. Open access has opened a wide variety of application for gas storage, particularly for marketers which can now exploit price arbitrage opportunities. Any storage capacity would be priced at cost-based pricing, unless the provider can demonstrate to FERC that it lacks market power, in which case it may be allowed to price at market-based rates to gain market share. FERC defines market power as "..the ability of a seller profitably to maintain prices above competitive levels for a significant period of time.". The underlying pricing structure for storage has discouraged development in the gas storage sector, which has not seen many new storage facilities constructed, besides current ones being expanded. In 2005, FERC announced a new Order 678 targeted particularly to gas storage. This rule is intended to stimulate the development of new gas storage facility in the ultimate goal of reducing natural gas price volatility. Commission Chairman Joseph T. Kelliher observed: "Since 1988, natural gas demand in the United States has risen 24 percent. Over the same period, gas storage capacity has increased only 1.4 percent. While construction of storage capacity has lagged behind the demand for natural gas, we have seen record levels of price volatility. This suggests that current storage capacity is inadequate. Further, this year, what storage capacity exists may be full far earlier than in any previous year. According to some analysts, that raises the prospect that some domestic gas production may be shut-in. Our final rule should help reduce price volatility and expand storage capacity." This ruling aims at opening up two approaches for developers of natural gas storage, to be able to charge market-based rates. The first one is the redefinition of the relevant product market for storage that includes alternatives for storage such as available pipeline capacity, local gas production and LNG terminals. The second approach aims at implementing section 312 of the Energy Policy Act. It would allow an applicant to request authority to charge "market-based rates even if a lack of market power has not been demonstrated, in circumstances where market-based rates are in the public interest and necessary to encourage the construction of storage capacity in the area needing storage service and that customers are adequately protected," the Commission said. It is expected that this new order will entice developers, especially independent storage operators, to develop new facilities in the near future.

## Iran

The Underground Gas storage Company, as a subsidiary of the National Iranian Gas company, was established in 2007 with the following objectives:

1. Organize, speed up and continue all the projects already in progress by the Natural Gas Storage Management

2. Implementation of new projects and plans

3. Execution of required studies with the purpose of locating new feasible gas storage potentials in different parts of the country which will turn into new plans and projects after approval. 6/30/2009.

## Sarajeh Reservoir

Sarajeh reservoir is depleted hydrocarbon reservoirs storage capacity of 3/3 billion cubic meters of natural gas, the company said. Storage capacity of the project to remove gas stored in the reservoir during the third phase follows considering are. - Phase 1: 10 million cubic meters of gas per day - Phase 2: 20 million cubic meters of gas per day - Phase 3: 30 million cubic meters of gas per day. The project started at the end of 86 years from the date 10/17/85, 16% and 54% of the initial installation of the goods has grown. After 87 years at the end of the company's natural gas storage development project, the installation of any goods supplied to 84% is reached. Electric power supply operation by the end of the project has not improved in 86 years at the end of year 87 5/96% is reached. The gas injection project design, installation and commissioning of gas injection compressors 87 till about 3/9% has been developed.

## Shurijeh Gas Storage Tank Project

The gas storage tank project in Shurijeh. Tank capacity approx eight quarters and the ability to produce 40 billion cubic meter. Gas in there and execute the following phases of the project have been considered: - The first phase of 10 AD. CE. M.. The injection of 20 m. CE. CE. - Phase II 20 m. M.. M.. R. injection and 40 mm.

## Canada

In Alberta, gas storage rates are not regulated and providers negotiate rates with their customers on a contract-by-contract basis. However the Carbon facility which is owned by ATCO gas is regulated, since ATCO is a utility company. Therefore, ATCO Gas has to charge cost-based rates for its customers, and can market any additional capacities at market-based rates. In Ontario, gas storage is regulated by the Ontario Energy Board. Currently all the available storage is owned by vertically integrated utilities. The utility companies have to price their storage capacity sold to their customers at cost-based rates, but can market any remaining capacity at market-based rates. Storage developed by independent storage developers can charge market-based rates. In British Columbia, gas storage is not regulated. All available storage capacity is marketed at market-based rates.

## United Kingdom

The regulation of gas storage, transportation and sale is overseen by Ofgem (a government regulator). This has been the case since the gas industry was privatised in 1986. Most forms of gas storage were owned by Transco (now part of National Grid plc), however the national network has now largely been broken down into regional networks, owned by different companies, they are however all still answerable to Ofgem.

## Storage Economics

### Storage Development Cost

As with all infrastructural investments in the energy sector, developing storage facilities is capital intensive. Investors usually use the return on investment as a financial measure for the viability of such projects. It has been estimated that investors require a rate or return between 12 percent to 15 percent for regulated projects, and close to 20 percent for unregulated projects. The higher expected return from unregulated projects is due to the higher perceived market risk. In addition significant expenses are accumulated during the planning and location of potential storage sites to determine its suitability, which further increases the risk.

The capital expenditure to build the facility mostly depends on the physical characteristics of the reservoir. First of all, the development cost of a storage facility largely depends on the type of the storage field. As a general rule of thumb, salt caverns are the most expensive to develop on a Bcf of Working Gas Capacity Basis. However one should keep in mind that because the gas in such facilities can be cycled repeatedly, on a Deliverability basis, they may be less costly. A Salt Cavern facility might cost anywhere from $10 million to $25 million/Bcf of working gas capacity. The wide price range is because of region difference which dictates the geological requirements. These factors include the amount of compressive horsepower required, the type of surface and the quality of the geologic structure to name a few. A depleted reservoir costs between $5 million to $6 million/Bcf of Working Gas Capacity. Finally another major cost incurred when building new storage facilities is that of base gas. The amount of base gas in a reservoir could be as high as 80% for aquifers making them very unattractive to develop when gas prices are high. On the other hand, salt caverns require the least amount of base gas. The high cost of base gas is what drives the expansion of current sites vs the development of new ones. This is because expansions require little addition to base gas.

The expected cash flows from such projects depend on a number of factors. These include the services the facility provides as well as the regulatory regime under which it operates. Facilities that operate primarily to take advantage of commodity arbitrage opportunities are expected to have different cash flow benefits than ones primarily used to ensure seasonal supply reliability. Rules set by regulators can on one hand restrict the profit made by storage facility owners or on the other hand guarantee profit, depending on the market model.

## Storage Valuation

To understand the economics of gas storage, it is crucial to be able to value it. Several approaches have been proposed. They include:

- Cost-of-service valuation

- Least-cost planning

- Seasonal valuation

- Option-based valuation

The different valuation modes co-exist in the real world and are not mutually exclusive. Buyers and sellers typically use a combination of the different prices to come up with the true value of storage. An example of the different valuations and the price they generate can be found in the table below.

| Cost of Storage Per Valuation Mode | |
|---|---|
| **Type** | **Dollars/mcf of working gas** |
| Median cost-of-service | $0.64 |

| | |
|---|---|
| Intrinsic value for Winter 05/06 as of August 2004 | $0.47-$0.62 |
| Least cost planning (depleted reservoir) | $0.70-$1.10 |
| Hypothetical cost-of-service of salt cavern | $2.93 |
| Intrinsic and extrinsic value of salt cavern (depleted reservoir) | $1.60-$1.90 |

## Cost-of-service Valuation

This valuation mode is typically used to value regulated storage, for instance storage operated by interstate pipeline companies. These companies are regulated by FERC. This pricing method allows the developers to recover their cost and an agreed upon return on investment. The regulatory body requires that the rates and tariffs are maintained and publicly published. The services provided by these companies include firm and interruptible storage as well as no-notice storage services. Usually, cost of service pricing is used for depleted reservoir facilities. If it is used to price, say salt cavern formations, the cost would be very high, due to the high cost of development of such facilities.

## Least-cost Planning

This valuation mode is typically used by local distribution companies (LDCs). It is based on pricing storage, according to the savings resulting from not having to resort to other more expensive options. This pricing mode depends on the consumer and their respective load profile/shape.

## Seasonal Valuation

The seasonal valuation of storage is also referred to as the *intrinsic value*. It is evaluated as the difference between the two prices in a pair of forward prices. The idea being that one can lock-in a forward spread, either physically or financially. For developers seeking to study the feasibility of building a storage facility, they would typically look at the long-term price spreads.

## Option-based Valuation

In addition to possessing an intrinsic value, storage may also have extrinsic value. Intrinsic valuation of storage does not take the cycling ability of high-deliverability storage. The extrinsic valuation reflects the fact that in such facilities, say salt cavern formations, a proportion of the space can be used more than once, thus increasing value. Such high-deliverability storage facility allows its user to respond to variations in demand/price within a season or during a given day rather than just seasonal variations as was the case with single cycle facilities.

## Effects of Natural Gas Prices on Storage

In general as we see in the graph below, high gas prices are typically associated to low storage periods. Usually when prices are high during the early months of the refill season (April–October), many users of storage adopt a wait and see attitude. They limit their gas intake in anticipation that the prices will drop before the heating season begins (November–March). However, when that decrease does not occur, they are forced to buy natural gas at high prices. This is particularly true for Local Distribution and other operators who rely on storage to meet the seasonal demand for their customers. On the other hand, other storage users, who use storage as a marketing tool (hedging or speculating) will hold off storing a lot of gas when the prices are high.

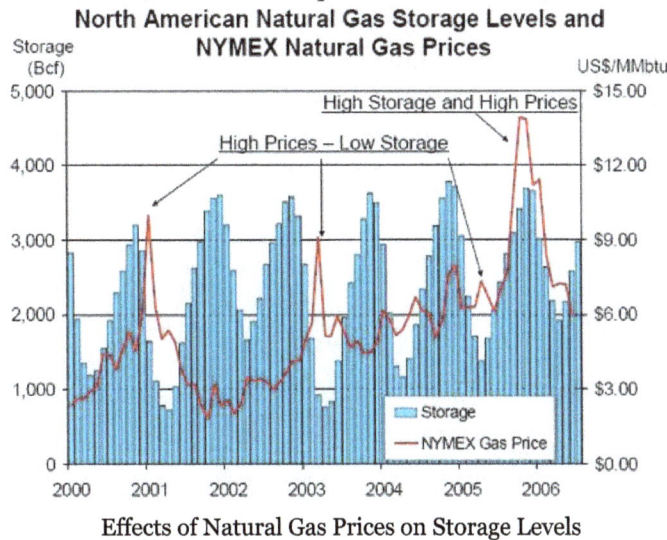

North American Natural Gas Storage Levels and NYMEX Natural Gas Prices

Effects of Natural Gas Prices on Storage Levels

## Future of Storage Technology

Research is being conducted on many fronts in the gas storage field to help identify new improved and more economical ways to store gas. Research being conducted by the US Department of Energy is showing that salt formations can be chilled allowing for more gas to be stored. This will reduce the size of the formation needed to be treated, and have salt extracted from it. This will lead to cheaper development costs for salt formation storage facility type. Another aspect being looked at, are other formations that may hold gas. These include hard rock formations such as granite, in areas where such formations exists and other types currently used for gas storage do not. In Sweden a new type of storage facility has been built, called "lined rock cavern". This storage facility consists of installing a steel tank in a cavern in the rock of a hill and surrounding it with concrete. Although the development cost of such facility is quite expensive, its ability to cycle gas multiple times compensates for it, similar to salt formation facilities. Finally, another research project sponsored by the Department of Energy, is that of hydrates. Hydrates are compounds formed when natural gas is frozen in the presence of water. The advantage being that as much as 181 standard cubic feet of natural gas could be stored in a single cubic foot of hydrate.

## References

- Marin S. Halper, James C. Ellenbogen (March 2006). Supercapacitors: A Brief Overview (PDF) (Technical report). MITRE Nanosystems Group. Retrieved January 20, 2014

- "Further demonstration of the VRLA-type UltraBattery® under medium-HEV duty and development of the flooded-type UltraBattery® for micro-HEV applications. Journal of Power Sources. 195: 1241. 2010. doi:10.1016/j.jpowsour.2009.08.080

- Chen, Y.; et., al. (2012). "1. Study on self-healing and lifetime characteristics of metallized-film capacitor under high electric field.". IEEE. 40. doi:10.1109/TPS.2012.2200699.

- Content, Thomas. Johnson Controls, UW Open Energy Storage Systems Test Lab In Madison, Milwaukee, Wisconsin: Milwaukee Journal Sentinel, May 5, 2014

- Hubler, A.; Osuagwu, O. (2010). "Digital quantum batteries: Energy and information storage in nanovacuum tube arrays". Complexity. 15. doi:10.1002/cplx.20306

- Aschenbrenner, Norbert. Test Plant For Automated Battery Production, Physics.org website, May 6, 2014. Retrieved May 8, 2014

- Belkin, Andrey; et., al. (2017). "Recovery of Alumina Nanocapacitors after High Voltage Breakdown". Sci.Rep. doi:10.1038/s41598-017-01007-9

# Accumulator and its Types

Accumulator stores, retrieves and releases energy. Various types of accumulators include steam accumulators, rechargeable batteries, flow batteries, etc. The chapter serves as a source to understand the major categories related to accumulators.

## Accumulator (Energy)

An accumulator is an energy storage device: a device which accepts energy, stores energy, and releases energy as needed. Some accumulators accept energy at a low rate (low power) over a long time interval and deliver the energy at a high rate (high power) over a short time interval. Some accumulators accept energy at a high rate over a short time interval and deliver the energy at a low rate over longer time interval. Some accumulators typically accept and release energy at comparable rates. Various devices can store thermal energy, mechanical energy, and electrical energy. Energy is usually accepted and delivered in the same form. Some devices store a different form of energy than what they receive and deliver performing energy conversion on the way in and on the way out.

Examples of accumulators include steam accumulators, mainsprings, flywheel energy storage, hydraulic accumulators, rechargeable batteries, capacitors, compensated pulsed alternators (compulsators), and pumped-storage hydroelectric plants.

In general usage in an electrical context, the word *accumulator* normally refers to a lead–acid battery.

The London Tower Bridge is operated via an accumulator. The original raising mechanism was powered by pressurised water stored in several hydraulic accumulators. In 1974, the original operating mechanism was largely replaced by a new electro-hydraulic drive system.

## Hydraulic Accumulator

A hydraulic accumulator is a pressure storage reservoir in which a non-compressible hydraulic fluid is held under pressure that is applied by an external source. The external source can be a spring, a raised weight, or a compressed gas. An accumulator enables a hydraulic system to cope with extremes of demand using a less powerful pump, to respond more quickly to a temporary demand, and to smooth out pulsations. It is a type of energy storage device.

Compressed gas accumulators, also called hydro-pneumatic accumulators, are by far the most common type.

## Types of Accumulators

### Towers

Grimsby Dock Tower

The first accumulators for Armstrong's hydraulic dock machinery were simple raised water towers. Water was pumped to a tank at the top of these towers by steam pumps. When dock machinery required hydraulic power, the hydrostatic head of the water's height above ground provided the necessary pressure.

These simple towers were extremely tall. One of the best known, Grimsby Dock Tower opened in 1852, is 300 feet (91 m) tall. The size of these towers made them expensive to construct. These simple tower accumulators were constructed for less than a decade. Around the same time, John Fowler was working on the construction of the ferry quay at nearby New Holland but could not use similar hydraulic power as the poor ground conditions did not permit a tall accumulator tower to be built. By the time Grimsby was opened, it was already obsolete as Armstrong had developed the more complex, but much smaller, weighted accumulator for use at New Holland. In 1892 the original Grimsby tower's function was replaced, on Fowler's advice, by a smaller weighted accumulator on an adjacent dock, although the tower remains to this day as a well-known landmark.

Another surviving tower is adjacent to East Float in Birkenhead, England.

### Raised Weight

Hydraulic engine house, Bristol Harbour

A raised weight accumulator consists of a vertical cylinder containing fluid connected to the hydraulic line. The cylinder is closed by a piston on which a series of weights are placed that exert a downward force on the piston and thereby pressurizes the fluid in the cylinder. In contrast to compressed gas and spring accumulators, this type delivers a nearly constant pressure, regardless of the volume of fluid in the cylinder, until it is empty. (The pressure will decline somewhat as the cylinder is emptied due to the decline in weight of the remaining fluid.)

A working example of this type of accumulator may be found at the hydraulic engine house, Bristol Harbour. The original 1887 accumulator is in place in its tower, an external accumulator was added in 1954 and this system was used until 2010 to power the Cumberland Basin (Bristol) lock gates. The water is pumped from the harbour into a header tank and then fed by gravity to the pumps. The working pressure is 750 psi (5.2 MPa, or 52 bar) which was used to power the cranes, bridges and locks of Bristol Harbour.

The original operating mechanism of Tower Bridge, London, also used this type of accumulator. Although no longer in use, two of the six accumulators may still be seen *in situ* in the bridge's museum.

Regent's Canal Dock, now named Limehouse Basin has the remains of a hydraulic accumulator, dating from 1869, a fragment of the oldest remaining such facility in the world, the second at the dock, which was installed later than that at Poplar Dock, originally listed incorrectly as a signalling cabin for the London and Blackwall Railway, when correctly identified, it was restored as a tourist attraction by the now defunct London Docklands Development Corporation. Now owned by the British Waterways Board, it is open for large groups on application to the Dockmaster's Office at the basin and on both the afternoons of London Open House Weekend, held on the third weekend of September each year.

London had an extensive public hydraulic power system from the mid-nineteenth century finally closing in the 1970s with 5 hydraulic power stations, operated by the London Hydraulic Power Company. Railway goods yards and docks often had their own separate system.

## Air-filled Accumulator

Steam fire engine, with vertical copper accumulator

A simple form of accumulator is an enclosed volume, filled with air. A vertical section of pipe, often enlarged diameter, may be enough and fills itself with air, trapped as the pipework fills.

Such accumulators do not have enough capacity to store power for long periods, but they can act as a buffer to absorb fluctuations in pressure. They are used to smooth out the delivery from piston pumps. Another use is as a shock absorber to damp out water hammer.

## Compressed Gas (or Gas-charged) Closed Accumulator

Piston accumulator

Citroën XM engine bay, showing two of Citroën's distinctive green spherical accumulators, used for the Hydropneumatic suspension system.

A compressed gas accumulator consists of a cylinder with two chambers that are separated by an elastic diaphragm, a totally enclosed bladder, or a floating piston. One chamber contains hydraulic fluid and is connected to the hydraulic line. The other chamber contains an inert gas under pressure (typically nitrogen) that provides the compressive force on the hydraulic fluid. Inert gas is used because oxygen and oil can form an explosive mixture when combined under high pressure. As the volume of the compressed gas changes, the pressure of the gas (and the pressure on the fluid) changes inversely.

It is possible to increase the gas volume of the accumulator by coupling a gas bottle to the gas side of the accumulator. This is mainly done since a gas bottle normally is cheaper to produce than an accumulator.

A compressed gas accumulator was invented by Sir. Jean Mercier, for use in variable-pitch propellers.

## Spring Type

A spring type accumulator is similar in operation to the gas-charged accumulator above, except that a heavy spring (or springs) is used to provide the compressive force. According to Hooke's law the magnitude of the force exerted by a spring is linearly proportional to its change of length. Therefore, as the spring compresses, the force it exerts on the fluid is increased linearly.

## Metal Bellows Type

The metal bellows accumulators function similarly to the compressed gas type, except the elastic diaphragm or floating piston is replaced by a hermetically sealed welded metal bellows. Fluid may be internal or external to the bellows. The advantages to the metal bellows type include exceptionally low spring rate, allowing the gas charge to do all the work with little change in pressure from full to empty, and a long stroke relative to solid (empty) height, which gives maximum storage volume for a given container size. The welded metal bellows accumulator provides an exceptionally high level of accumulator performance, and can be produced with a broad spectrum of alloys resulting in a broad range of fluid compatibility. Another advantage to this type is that it does not face issues with high pressure operation, thus allowing more energy storage capacity.

## Functioning of an Accumulator

In modern, often mobile, hydraulic systems the preferred item is a gas charged accumulator, but simple systems may be spring-loaded. There may be more than one accumulator in a system. The exact type and placement of each may be a compromise due to its effects and the costs of manufacture.

An accumulator is placed close to the pump with a non-return valve preventing flow back to the pump. In the case of piston-type pumps this accumulator is placed in the ideal location to absorb pulsations of energy from the multi-piston pump. It also helps protect the system from fluid hammer. This protects system components, particularly pipework, from both potentially destructive forces.

An additional benefit is the additional energy that can be stored while the pump is subject to low demand. The designer can use a smaller-capacity pump. The large excursions of system components, such as landing gear on a large aircraft, that require a considerable volume of fluid can also benefit from one or more accumulators. These are often placed close to the demand to help overcome restrictions and drag from long pipework runs. The outflow of energy from a discharging accumulator is much greater, for a short time, than even large pumps could generate.

An accumulator can maintain the pressure in a system for periods when there are slight leaks without the pump being cycled on and off constantly. When temperature changes cause pressure excursions the accumulator helps absorb them. Its size helps absorb fluid that might otherwise be locked in a small fixed system with no room for expansion due to valve arrangement.

The gas precharge in an accumulator is set so that the separating bladder, diaphragm or piston does not reach or strike either end of the operating cylinder. The design precharge normally ensures that the moving parts do not foul the ends or block fluid passages. Poor maintenance of precharge can destroy an operating accumulator. A properly designed and maintained accumulator should operate trouble-free for years.

# Steam Accumulator

A steam accumulator is an insulated steel pressure tank containing hot water and steam under pressure. It is a type of energy storage device. It can be used to smooth out peaks and troughs in demand for steam. Steam accumulators may take on a significance for energy storage in solar thermal energy projects. An example is the PS10 solar power tower plant near Seville, Spain and one planned for the "solar steam train" project in Sacramento, California.

District heating steam accumulator tower on the Churchill Gardens Estate, Pimlico, London, United Kingdom. This plant once used waste heat piped from Battersea Power Station on the opposite side of the River Thames. (January 2006)

## History

It was invented in 1874 by the Scottish engineer Andrew Betts Brown.

## Charge

The tank is about half-filled with cold water and steam is blown in from a boiler via a perforated pipe near the bottom of the drum. Some of the steam condenses and heats the water. The remainder fills the space above the water level. When the accumulator is fully charged the condensed steam will have raised the water level in the drum to about three-quarters full and the temperature and pressure will also have risen.

## Discharge

Steam can be drawn off as required, either for driving a steam turbine or for process purposes (e.g. in chemical engineering), by opening a steam valve on top of the drum. The pressure in the drum will fall but the reduced pressure causes more water to boil and the accumulator can go on supplying steam (while gradually reducing pressure and temperature) for some time before it has to be re-charged.

## Pressure and Temperature

This steam table shows the relationship between pressure and temperature in a boiler or steam accumulator:

| Gauge Pressure, PSI (bar) | Absolute Pressure, PSI (bar) | Temperature, °F (°C) |
|---|---|---|
| 0 (0) | 15 (1) | 212 (100) |
| 35 (2.4) | 50 (3.4) | 281 (138) |
| 85 (5.9) | 100 (6.9) | 328 (164) |
| 135 (9.3) | 150 (10.3) | 358 (181) |
| 185 (12.8) | 200 (13.8) | 382 (194) |
| 235 (16.2) | 250 (17.2) | 401 (205) |

## Abbreviations and Notes

- PSI = pounds-force per square inch

- Absolute pressure = gauge pressure + atmospheric pressure

- °C = degrees Celsius

- °F = degrees Fahrenheit

# Home Energy Storage

Home energy storage devices store electricity locally, for later consumption. At their heart are batteries, typically lithium-ion or lead-acid, and intelligent software. An energy storage technology, they are downstream relatives of battery-based grid energy storage and support the concept of distributed generation. When paired with on-site generation, they can support an off-the-grid lifestyle.

## Operating Modes

### On-site Generation

The stored energy commonly originates from on-site solar photovoltaic panels, generated during daylight hours, and the stored electricity consumed after sundown, when domestic energy demand peaks in homes unoccupied during the day.

Electric vehicles (EVs) used during weekdays, needing recharging overnight, are a good fit with home energy storage in homes with solar panels and low daylight-hour electrical consumption. EV manufacturers Tesla, Mercedes-Benz, BMW, Nissan and BYD market own-brand home energy storage devices to their customers, with Tesla's Powerwall enjoying significant media exposure.

### Differential Tariffs and Smart Meters

The units can also be programmed to exploit a differential tariff, that provide lower priced energy during hours of low demand - seven hours from 12:30am in the case of Britain's Economy 7 tariff - for consumption when prices are higher.

Smart tariffs, stemming from the increasing prevalence of smart meters, will increasingly be paired with home energy storage devices to exploit low off-peak prices, and avoid higher-priced energy at times of peak demand.

## Advantages

### Overcoming Grid Losses

Transmission of electrical power from power stations to population centres is inherently inefficient, due to transmission losses in electrical grids, particularly within power-hungry dense conurbations where power stations are harder to site. By allowing a greater proportion of on-site generated electricity to be consumed on-site, rather than exported to the energy grid, home energy storage devices can reduce the inefficiencies of grid transport.

### Energy Grid Support

Home energy storage devices, when connected to a server via the internet, can theoretically be ordered to provide very short-term services to the energy grid:-

- Reduced peak hour demand stress - provision of short-term demand response during periods of peak demand reducing the need to inefficiently standing up of short generation assets like diesel generators.

- Frequency correction - the provision of ultra short-term corrections, to keep mains frequency within the tolerances required by regulators (e.g. 50hz or 60hz +/- n%).

### Reduced Reliance on Fossil Fuels

Due to the above efficiencies, and their ability to boost the amount of solar energy consumed on-site, the devices reduce the amount of power generated using fossil fuels, namely natural gas, coal, oil and diesel.

## Disadvantages

### Environmental Impact of Batteries

Lithium-ion batteries, a popular choice due to their relatively high charge cycle and lack of memory effect, are difficult to recycle.

Lead-acid batteries are relatively easier to recycle and, due to the high resale value of the lead, 99% of those sold in the US get recycled. They have much shorter useful lives than a lithium-ion battery of a similar capacity, due to having a lower charge cycle, narrowing the environmental-impact gap. In addition, lead is a toxic heavy metal and the sulphuric acid in the electrolyte has a high environmental impact.

### Second Life for Ev Batteries

To offset the environmental impact of batteries, some manufacturers extend the useful life of used batteries taken from electric vehicles at the point where the cells won't sufficiently hold charge.

Though considered end of life for electric vehicles, the batteries will function satisfactorily in home energy storage devices. Manufacturers supporting this include Nissan, BMW and Powervault.

## Salt Water Batteries

Home Energy Storage devices can be paired with salt water batteries, which have a lower environmental impact due to their lack of toxic heavy metal and ease of recyclability.

## Tesla Powerwall

The Powerwall and Powerpack are rechargeable lithium-ion battery stationary energy storage products manufactured by Tesla, Inc. The Powerwall is intended to be used for home energy storage and stores electricity for solar self-consumption, time of use load shifting, backup power, and off-the-grid use. The larger Powerpack is intended for commercial or electric utility grid use and can be used for peak shaving, load shifting, backup power, demand response, microgrids, renewable power integration, frequency regulation, and voltage control.

Announced in 2015, with a pilot demonstration of 500 units built and installed during 2015, production of the product was initially at the Tesla Fremont factory before being moved to the under construction Gigafactory 1 in Nevada. The second generation of both products were announced in October 2016.

## History

Tesla started development in 2012, installing prototypes at selected industrial customers. In some cases, PowerPacks have saved 20% of the electrical bill. The Powerwall was originally announced at the April 30, 2015 product launch with power output of 2 kW steady and 3.3 kW peak, but Musk said at the June 2015 Tesla shareholders meeting that this would be more than doubled to 5 kW steady with 7 kW peak, with no increase in price. He also announced that Powerwall deliveries would be prioritized to partners who minimize the cost to the end user, with a Powerwall installation price of US$500.

When originally announced in 2015, there were to be two models of Powerwall delivered: 10 kWh capacity for backup applications and 7 kWh capacity for daily cycle applications. But by March 2016, Tesla had "quietly removed all references to its 10-kilowatt-hour residential battery from the Powerwall website, as well as the company's press kit. The company's smaller battery designed for daily cycling is all that remains." The 10 kWh battery as originally announced has a nickel-cobalt-aluminum cathode, like the Tesla Model S, which was projected to be used as a backup/uninterruptible power supply, and had a projected cycle life of 1000–1500 cycles.

In October 2016, Tesla announced that nearly 300 MWh of Tesla batteries had been deployed in 18 countries. The Powerwall 2 was unveiled in October 2016 at Universal Studios' Colonial Street, Los Angeles, backlot street set and is designed to work with the solar panel roof tiles to be produced by SolarCity.

## Powerwall Specifications

Since March 2016, there was only a single model: the 6.4 kWh version for daily cycle applications, before the Powerwall 2 was introduced:

| Model | Tech-nology | Price (US$) | Capaci-ty (kWh) | Wh per US$ | US$ per kWh | Power | Operat-ing temp. | Weight | Dimen-sions, H × W × D | Cycles (during warranty) | US$ per warranted kWh |
|---|---|---|---|---|---|---|---|---|---|---|---|
| Power-wall 1 | Lithi-um-ion | US$3,000 | 6.4 | 2.13 | 469 | 7 kW peak; 5 kW continu-ous | −4 to 110 °F (−20 to 43 °C) | 214 lb (97 kg) | 51.3 in × 34 in × 7.2 in (130 cm × 86 cm × 18 cm) | 5,000 | |
| Power-wall 2 | Lithi-um-ion | US$5,500 | 13.5 | 2.46 | 407 | 7 kW peak; 5 kW continu-ous | −4 to 122 °F (−20 to 50 °C) | 264.4 lb (119.9 kg) | 44 in × 29 in × 5.5 in (112 cm × 74 cm × 14 cm) | | ~0.17 |

## Powerpack Specifications

| Model | Technology | Capacity (kWh) | Wh per US$ | US$ per kWh | Operating temp. | Weight | Dimensions, H × W × D |
|---|---|---|---|---|---|---|---|
| Powerpack 1 | Lithium-ion | 100 | 2.13 | 470 | - | - | 218.5 in × 82.2 in × 130.8 in (555 cm × 209 cm × 332 cm) |
| Powerpack 2 | Lithium-ion | 200 | 2.51 | 398 | −22 to 122 °F (−30 to 50 °C) | 3,575 lb (1,622 kg) | 218.5 in × 82.2 in × 130.8 in (555 cm × 209 cm × 332 cm) |

## Example of Powerpack Installation

Tesla installed a grid storage facility for the Southern California Edison with a capacity of 80 MWh at a power of 20 MW between September 2016 and December 2016. This means that the storage unit (1/2017) is currently one of the largest accumulator batteries on the market. Tesla installed 400 Powerpack-2 modules at the Mira Loma transformer station in California. The memory serves to store energy at a low network load and then to feed this energy back into the grid at peak load. Prior to this, gas-fired power stations were used.

## Versions

The first generation Powerwall has a 6.4 kWh capacity for daily cycle applications. For families with larger energy needs, multiple powerwalls can be connected to expand the capacity even higher. A previously announced 10 kWh capacity model designed for backup power purposes was quietly discontinued in March 2016, as the 6.4 kWh version can also be configured to act as backup power.

The Powerpack is a bigger unit with 100 kWh (first generation) and 210 kWh (2nd generation) of storage for commercial and utility grid use. In order to meet the variety of energy needs in industry, "Powerpack is infinitely scalable", said Elon Musk. Tesla's objective is to "fundamentally change the way the world uses energy" by "fostering a clean energy ecosystem and helping wean the world off fossil fuels" using backup energy storage for renewable energy. The Powerpack 2 has 200 kWh of storage, probably using the 2170 cell by the end of 2016.

## Technology

The Powerwall is optimized for daily cycling, such as for load shifting. Tesla uses proprietary technology for packaging and cooling the cells in packs with liquid coolant. Elon Musk, the chairman, CEO and product architect of the Tesla company, promised not to start patent infringement lawsuits against anyone who, in good faith, used Tesla's technology for Powerwalls as he had promised with Tesla cars.

The daily cycle 7 kWh PW1 battery uses nickel-manganese-cobalt chemistry and can be cycled 5,000 times before warranty expiration. The Tesla Powerwall has a 92.5% round trip efficiency when charged or discharged by a 400–450 V system at 2 kW with a temperature of 77 °F (25 °C), and when the product is brand new. Age of the product, temperatures above or below 77 °F (25 °C), and charge rates or discharge rates above 2 kW would lower this efficiency number, decreasing the system performance.

First generation Powerwalls includes a DC-to-DC converter to sit between a home's existing solar panels, and the home's existing DC to AC inverter. If the existing inverter is not storage-ready, one must be purchased. The second generation Powerwall incorporates a DC-to-AC inverter of Tesla's own design. Production of the 2170 cell for the PW2 and PP2 began at Gigafactory in January 2017.

The National Fire Protection Association conducted two worst-case scenario tests in 2016, igniting Powerpacks to initiate thermal run-off. The design contained damage to the structures.

## Market

### Powerwall

The Powerwall was unveiled on April 30, 2015, with a 7 kWh Powerwall model that would retail for US$3,000 and a 10 kWh model at US$3,500. Shipments of 500 pilot units were planned to begin in the late summer of 2015. Musk indicated that he believed the low Tesla price would cause other storage producers to follow. Before the April 30, 2015, unveiling, some existing solar-panel users participated in a demonstration program and paid up to US$13,000 for a 10 or 15 kWh Tesla battery.

As of May 2015, Powerwalls were sold to companies including SolarCity and OUXO Energy for installation. SolarCity was running a pilot project in 500 California houses, using 10 kWh battery packs. A market overview calculates Powerwall 2 at 0.23 Australian dollars per warranted kWh.

## Volume Tendency

As of May 2015, Tesla Powerwall had already sold out through to the middle of 2016. Reservations within the first few weeks were over 50,000 units for the Powerwall (US$179 million), and 25,000 units for the Powerpack (US$625 million), therefore combined orders of US$800 million.

During the first quarter of 2016, Tesla delivered over 25 MWh of energy storage to customers on four continents. Over 2,500 Powerwalls and nearly 100 Powerpacks were delivered in North America, Asia, Europe, and Africa. The first Powerwall in Portugal has been sold by OUXO Energy. As of October 2016, nearly 300 MWh of Tesla batteries had been deployed worldwide.

## Powerpack

At the announcement, a larger battery called Powerpack—storing 100 kWh of electrical energy—was projected to be available for industrial consumers, reaching a price point of $250/kWh. The Powerpack was projected to comprise the majority of stationary storage production at Gigafactory 1 while Powerwall would play a smaller part, giving Tesla a profit margin of 20 percent.

## Price Tendency

In September 2016, Tesla priced the Powerpack at $445/kWh, and a system with 200 kWh of energy and 100 kW of peak power was the cheapest available priced at $145,100. A bi-directional 250 kW inverter costs $52,500. By October 2016, a limited system of Powerpack 2 cost $398/kWh.

## Volume Tendency

Musk predicted in 2016 that the utility power will need to increase to supply more electric vehicles, eventually reaching an equilibrium with about 1/3 of power coming from distributed energy and 2/3 from utilities. Battery storage is one of the ways to mitigate the increasing duck curve, particularly in California.

## Return on Investment Calculations

A May 2015 article in Forbes magazine calculated that using a Tesla Powerwall 1 model combined with solar panels in a home would cost 30 cents/kWh for electricity if a home remains connected to the grid (the article acknowledges that the Tesla battery could make economic sense in applications that are entirely off-grid). US consumers got electricity from the power grid for 12.5 cents/kWh on average. The article concluded the "...Tesla's Powerwall Is Just Another Toy For Rich Green People." Bloomberg and Catalytic Engineering magazines also agreed that the Tesla system was most useful in places where electricity prices are high.

There are however a number of such locations, including Hawaii and other remote islands that generate electricity with shipped-in or flown-in fuels. Residential California PG&E customers pay as much as 40 cents/kWh if they reach Tier 3 in electrical usage, which is quite easy to do if you have a swimming pool pump, even an energy-efficient pump. Arctic and sub-Arctic locations with high energy prices cannot generate sufficient solar energy in the winter due to little or no sunlight.

The Swiss bank UBS said that the Powerwall makes sense in Australia and Germany where electricity is very costly but solar panels are well distributed.

As of November 2016 cost of installation for one Powerwall 2 starts at AU$8750 in Australia or US$1,600 in USA Powerwall 1 price (Feb 2017)- US$3000 (sized residential use). Powerwall 2 price (Feb 2017) - US$5500 (larger capacity, but still residential-size, will run many homes for 2 or 3 days without outside power if fully charged).

## Competition

### Home

Energy technology company Enphase Energy, based in California, has announced it will release its lithium iron phosphate battery as part of a complete alternating current Enphase Home Energy Solution starting in Winter 2016 in Australia and New Zealand with Genesis Energy conducting trials. The system, which includes monitoring and control of solar generation, home energy consumption and battery storage, will be sold at wholesale through solar distributors, who sell to solar installers. Enphase's modular 'building block' batteries are more efficient than the Tesla Powerwall (96% compared to Tesla's 92% round-trip efficiency). The Enphase AC Battery also includes an inverter inside the casing, and works with all existing solar systems, or alternatively in homes without solar. Lithium iron phosphate batteries are known to be the most stable and safe of the various lithium batteries.

Mercedes-Benz / Daimler AG announced in June 2015 that they would be selling batteries for domestic or commercial use by the end of 2015. These would compete against the Tesla Powerwall and would be marketed by *Deutsche Accumotive*, the Daimler subsidiary that produces the Lithium-ion battery that Mercedes uses in its electric and hybrid cars. Several other solutions are available.

### Industrial

BYD's energy storage system is another competitor of Tesla's Powerpack. UC San Diego installed this system which has 5 megawatt-hour (MWh) capacity—enough to power 2,500 homes—in September 2014. BYD is a large supplier of rechargeable batteries, and is also known for its leading position in electric buses.

Sonnen and AutoGrid collaborates on combining house batteries into a large scale utility-level grid storage system. Eos claimed a battery price of $160/kWh in 2017, before the cost of integration by Siemens.

## Flow Battery

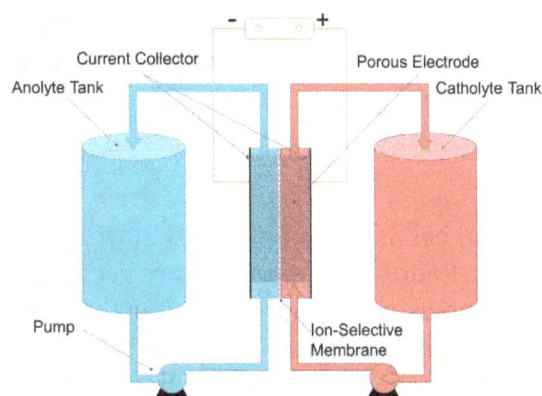

A typical flow battery consists of two tanks of liquids which are pumped past
a membrane held between two electrodes

A flow battery, or redox flow battery (after *reduction–oxidation*), is a type of electrochemical cell where chemical energy is provided by two chemical components dissolved in liquids contained within the system and separated by a membrane. Ion exchange (accompanied by flow of electric current) occurs through the membrane while both liquids circulate in their own respective space. Cell voltage is chemically determined by the Nernst equation and ranges, in practical applications, from 1.0 to 2.2 volts.

A flow battery may be used like a fuel cell (where the spent fuel is extracted and new fuel is added to the system) or like a rechargeable battery (where an electric power source drives regeneration of the fuel). While it has technical advantages over conventional rechargeables, such as potentially separable liquid tanks and near unlimited longevity, current implementations are comparatively less powerful and require more sophisticated electronics.

The energy capacity is a function of the electrolyte volume (amount of liquid electrolyte) and the power a function of the surface area of the electrodes.

## Construction Principle

A flow battery is a rechargeable fuel cell in which an electrolyte containing one or more dissolved electroactive elements flows through an electrochemical cell that reversibly converts chemical energy directly to electricity (electroactive elements are "elements in solution that can take part in an electrode reaction or that can be adsorbed on the electrode"). Additional electrolyte is stored externally, generally in tanks, and is usually pumped through the cell (or cells) of the reactor, although gravity feed systems are also known. Flow batteries can be rapidly "recharged" by replacing the electrolyte liquid (in a similar way to refilling fuel tanks for internal combustion engines) while simultaneously recovering the spent material for re-energization.

In other words, a flow battery is just like an electrochemical cell, with the exception that the ionic solution (electrolyte) is not stored in the cell around the electrodes. Rather, the ionic solution is stored outside of the cell, and can be fed into the cell in order to generate electricity. The total amount of electricity that can be generated depends on the size of the storage tanks.

Flow batteries are governed by the design principles established by electrochemical engineering.

## Types

Various types of flow cells (batteries) have been developed, including redox, hybrid and membraneless. The fundamental difference between conventional batteries and flow cells is that energy is stored not as the electrode material in conventional batteries but as the electrolyte in flow cells.

## Redox

The redox (reduction–oxidation) cell is a reversible cell in which electrochemical components are dissolved in the electrolyte. Redox flow batteries are rechargeable (secondary cells). Because they employ heterogeneous electron transfer rather than solid-state diffusion or intercalation they are more appropriately called fuel cells than batteries. In industrial practice, fuel cells are usually, and unnecessarily, considered to be primary cells, such as the $H_2/O_2$ system. The unitized regenerative fuel cell on NASA's Helios Prototype is another reversible fuel cell. The European Patent Organisation classifies redox flow cells (H01M8/18C4) as a sub-class of regenerative fuel cells (H01M8/18). Examples of redox flow batteries are the Vanadium redox flow battery, polysulfide bromide battery (Regenesys), and uranium redox flow battery. Redox fuel cells are less common commercially although many systems have been proposed.

A prototype zinc-polyiodide flow battery has been demonstrated with an energy density of 167 Wh/l (watt-hours per liter). Older zinc-bromide cells reach 70 Wh/l. For comparison, lithium iron phosphate batteries store 233 Wh/l. The zinc-polyiodide battery is claimed to be safer than other flow batteries given its absence of acidic electrolytes, nonflammability and operating range of −4 to 122 °F (−20 to 50 °C) that does not require extensive cooling circuitry, which would add weight and occupy space. One unresolved issue is zinc build-up on the negative electrode that permeated the membrane, reducing efficiency. Because of the Zn dendrite formation, the Zn-halide batteries cannot operate at high current density (> 20 mA/cm$^2$) and thus have limited power density. Adding alcohol to the electrolyte of the ZnI battery can slightly control the problem.

When the battery is fully discharged, both tanks hold the same electrolyte solution: a mixture of positively charged zinc ions (Zn2+ ) and negatively charged iodide ion, I-. When charged, one tank holds another negative ion, polyiodide, I3-. The battery produces power by pumping liquid from external tanks into the battery's stack area where the liquids are mixed. Inside the stack, zinc ions pass through a selective membrane and change into metallic zinc on the stack's negative side.

Traditional flow battery chemistries have both low specific energy (which makes them too heavy for fully electric vehicles) and low specific power (which makes them too expensive for stationary energy storage). However, recently a high areal power of 1.4 W/cm2 was demonstrated for hydrogen-bromine flow batteries, and a specific energy (530 Wh/kg at the tank level) was shown for hydrogen-bromate flow batteries.

One system uses organic polymers and a saline solution with a cellulose membrane. The prototype withstood 10000 charging cycles while retaining substantial capacity. The energy density was 10 Wh/l. Current density reached 100 milliamperes/cm2.

## Hybrid

The hybrid flow battery uses one or more electroactive components deposited as a solid layer. In this case, the electrochemical cell contains one battery electrode and one fuel cell electrode. This type is limited in energy by the surface area of the electrode. Hybrid flow batteries include the zinc-bromine, zinc–cerium and lead–acid flow batteries.

## Membraneless

A membraneless battery relies on laminar flow in which two liquids are pumped through a channel. They undergo electrochemical reactions to store or release energy. The solutions stream through in parallel, with little mixing. The flow naturally separates the liquids, eliminating the need for a membrane.

Membranes are often the most costly component and the least reliable components of batteries, as they can corrode with repeated exposure to certain reactants. The absence of a membrane enabled the use of a liquid bromine solution and hydrogen. This combination is problematic when membranes are used, because they form hydrobromic acid that can destroy the membrane. Both materials are available at low cost.

The design uses a small channel between two electrodes. Liquid bromine flows through the channel over a graphite cathode and hydrobromic acid flows under a porous anode. At the same time, hydrogen gas flows across the anode. The chemical reaction can be reversed to recharge the battery—a first for any membraneless design. One such membraneless flow battery published in August 2013 produced a maximum power density of 7950 W/m², three times as much power as other membraneless systems— and an order of magnitude higher than lithium-ion batteries.

Primus Power has developed patented technology in its zinc bromine flow battery, a type of redox flow battery, to eliminate the membrane or separator, which reduces costs and reduces failure rates. The Primus Power membraneless redox flow battery is working in installations in the United States and Asia with a second generation product announced 21 February 2017. Primus Power EnergyPod 2 is in production.

## Organic

Compared to traditional aqueous inorganic redox flow batteries such as vanadium redox flow batteries and Zn-Br2 batteries, which have been developed for decades, organic redox flow batteries emerged in 2009 and hold great promise to overcome major drawbacks preventing economical and extensive deployment of traditional inorganic redox flow batteries. The primary merit of organic redox flow batteries lies in the tunable redox properties of the redox-active components.

Organic redox flow batteries could be further classified into two categories: Aqueous Organic Redox Flow Batteries (AORFBs) and Non-aqueous Organic Redox Flow Batteries (NAORFBs). AORFBs use water as solvent for electrolyte materials while NAORFBs employ organic solvents to dissolve redox active materials. Depending on using one or two organic redox active electrolytes as anode and/or cathode, AORFBs and NAORFBs can be further divided into total organic systems and hybrid organic systems that use inorganic materials for anode or cathode. The proof of concept of AORFBs was demonstrated before NAORFBs. In larger-scale energy storage, AORFBs hold much greater potential than NAORFBs because of the former's lower cost, higher current and power performance, as well as safety advantages of aqueous over non-aqueous electrolytes. NAORFBs may find their place in limited special applications for their higher energy density than AORFBs though more challenges are to be overcome in terms of safety, cost of organic solvents, radical induced side reactions, electrolyte crossover and limited life time. The content below mainly covers the representative studies on AORFBs.

Quinones are the basis of some AORFBs. and 2010. In one study, 1,2-dihydrobenzoquinone-3,5-disulfonic acid (BQDS) and 1,4-dihydrobenzoquinone-2-sulfonic acid (BQS) were employed as cathodes, and conventional $Pb/PbSO_4$ was the anolyte in an acid AORFB. These first AORFBs are hybrid systems as they use organic redox active materials only for the cathode side. The quinones accepts two units of electrical charge, compared with one in conventional catholyte, implying that such a battery could store twice as much energy in a given volume.

9,10-Anthraquinone-2,7-disulphonic acid (AQDS), also a quinone, has been evaluated as well. AQDS undergoes rapid, reversible two-electron/two-proton reduction on a glassy carbon electrode in sulphuric acid. An aqueous flow battery with inexpensive carbon electrodes, combining the quinone/hydroquinone couple with the Br 2/Br− redox couple, yields a peak galvanic power density exceeding 6,000 W/m² at 13,000 A/m². Cycling showed >99 % storage capacity retention per cycle. Volumetric energy density was over 20 Wh/l. Anthraquinone-2-sulfonic acid and anthraquinone-2,6-disulfonic acid on the negative side and 1,2-dihydrobenzoquinone- 3,5-disulfonic acid on the positive side avoids the use of hazardous $Br_2$. The battery was claimed to last for 1,000 cycles without degradation although no official data were published. While this total organic system appears robust, it has a low cell voltage (ca. 0.55 V) and a low energy density (< 4 Wh/L).

Hydrobromic acid used as an electrolyte has been replaced with a far less toxic alkaline solution (1M KOH) and ferrocyanide. The higher pH is less corrosive, allowing the use of inexpensive polymer tanks. The increased electrical resistance in the membrane was compensated by increasing the voltage. The cell voltage was 1.2 V. The cell's efficiency exceeded 99%, while round-trip efficiency measured 84%. The battery has an expected lifetime of at least 1,000 cycles. Its theoretic energy density was 19 Wh per liter. Ferrocyanide's chemically stability in high pH KOH solution without forming Fe(OH)2 or Fe(OH)3 needs to be verified before scale-up.

Another organic AORFB has been demonstrated methyl viologen as anolyte and 4-hydroxy-2,2,6,6-tetramethylpiperidin-1-oxyl as catholyte, plus sodium chloride and a low-cost anion exchange membrane to enable charging and discharging. This MV/TEMPO system has the highest cell voltage, 1.25 V, and, possibly, lowest capital cost ($180/kWh) reported for AORFBs. The water-based liquid electrolytes were designed as a drop-in replacement for current systems without replacing existing infrastructure. A 600-milliwatt test battery was stable for 100 cycles with nearly 100 percent efficiency at current densities ranging from 20 to 100 mA per square centimeter, with optimal performance rated at 40-50 mA, at which about 70 percent of the battery's original voltage was retained. The significance of the research is that neutral AORFBs can be more environmentally friendly than acid or alkaline AORFBs while showing electrochemical performance comparable to corrosive acidic or alkaline RFBs. The MV/TEMPO AORFB has an energy density of 8.4 Wh/L with the limitation on the TEMPO side. The next step is to identify a high capacity catholyte to match MV (ca. 3.5 M solubility in water, 93.8 Ah/L).

One flow-battery concept is based on redox active, organic polymers employs viologen and TEMPO with dialysis membranes. The polymer-based redox-flow battery (pRFB) uses functionalized macromolecules (similar to acrylic glass or Styrofoam) being dissolved in water as active material for the anode as well as the cathode. Thereby, metals and strongly corrosive electrolytes – like vanadium salts in sulfuric acid – are avoided and simple dialysis membranes can be employed. The membrane, which separates the cathode and the anode of the flow cell, works like a strainer and is produced much more easily and at lower cost than conventional ion-selective membranes. It retains the big "spaghetti"-like polymer molecules, while allowing the small counterions to pass. The concept may solve the high cost of traditional Nafion membrane but the design and synthesis of redox active polymer with high solubility in water is not trivial.

## Metal Hydride

Proton flow batteries integrate a metal hydride storage electrode into a reversible proton exchange membrane (PEM) fuel cell. During charging, PFB combines hydrogen ions produced from splitting water with electrons and metal particles in one electrode of a fuel cell. The energy is stored in the form of a solid-state metal hydride. Discharge produces electricity and water when the process is reversed and the protons are combined with ambient oxygen. Metals less expensive than lithium can be used and provide greater energy density than lithium cells.

## Nano-network

Lithium–sulfur system arranged in a network of nanoparticles eliminates the requirement that charge moves in and out of particles that are in direct contact with a conducting plate. Instead, the nanoparticle network allows electricity to flow throughout the liquid. This allows more energy to be extracted.

## Semi-solid

In a semi-solid flow cell, the positive and negative electrodes are composed of particles suspended in a carrier liquid. The positive and negative suspensions are stored in separate tanks and pumped through separate pipes into a stack of adjacent reaction chambers, where they are separated by a barrier such as a thin, porous membrane. The approach combines the basic structure of aqueous-flow batteries, which use electrode material suspended in a liquid electrolyte, with the chemistry of lithium-ion batteries in both carbon-free suspensions and slurries with conductive carbon network. The carbon free semi-solid redox flow battery is also sometimes referred to as Solid Dispersion Redox Flow Battery. Dissolving a material changes its chemical behavior significantly. However, suspending bits of solid material preserves the solid's characteristics. The result is a viscous suspension that flows like molasses.

## Chemistries

A wide range of chemistries have been tried for flow batteries.

| Couple | Max. Cell Voltage (V) | Average Electrode Power Density (W/m²) | Average Fluid Energy Density (W·h/kg) | Cycles |
|---|---|---|---|---|
| Hydrogen–lithium bromate | 1.1 | 15,000 | 750 | |
| Hydrogen–lithium chlorate | 1.4 | 10,000 | 1400 | |
| Bromine-hydrogen | 1.07 | 7,950 | | |
| Iron–tin | 0.62 | <200 | | |
| Iron–titanium | 0.43 | <200 | | |
| Iron–chrome | 1.07 | <200 | | |
| Organic (2013) | 0.8 | 13000 | 21.4 Wh/L | 10 |
| Organic (2015) | 1.2 | | 7.1 Wh/L | 100 |
| MV-TEMPO | 1.25 | | 8.4 Wh/L | 100 |
| Vanadium–vanadium (sulphate) | 1.4 | ~800 | 25 Wh/L | |
| Vanadium–vanadium (bromide) | | | 50 Wh/L | 2000 |
| Sodium–bromine polysulfide | 1.54 | ~800 | | |
| Zinc–bromine | 1.85 | ~1,000 | 75 | |
| Lead–acid (methanesulfonate) | 1.82 | ~1,000 | | |
| Zinc–cerium (methanesulfonate) | 2.43 | <1,200–2,500 | | |

## Advantages and Disadvantages

Redox flow batteries, and to a lesser extent hybrid flow batteries, have the advantages of flexible layout (due to separation of the power and energy components), long cycle life (because there are no solid-to-solid phase transitions), quick response times, no need for "equalisation" charging (the overcharging of a battery to ensure all cells have an equal charge) and no harmful emissions. Some types also offer easy state-of-charge determination (through voltage dependence on charge), low maintenance and tolerance to overcharge/overdischarge. Compared to solid-state rechargeable batteries such as Li ion, RFBs, and ARFBs in particular, can operate at much higher current and power densities. These technical merits make redox flow batteries a well-suited option for large-scale energy storage.

On the negative side, the energy densities vary considerably but are, in general, lower compared to portable batteries, such as the Li-ion.

## Applications

Flow batteries are normally considered for relatively large (1 kWh – 10 MWh) stationary applications. These are for

- Load balancing – where the battery is connected to an electrical grid to store excess electrical power during off-peak hours and release electrical power during peak demand periods. The common problem limiting the use of most flow battery chemistries in this application is their low areal power (operating current density) which translates into a high cost of power.

- Storing energy from renewable sources such as wind or solar for discharge during periods of peak demand.

- Peak shaving, where spikes of demand are met by the battery.

- UPS, where the battery is used if the main power fails to provide an uninterrupted supply.

- Power conversion – because all cells share the same electrolyte/s. Therefore, the electrolyte/s may be charged using a given number of cells and discharged with a different number. Because the voltage of the battery is proportional to the number of cells used the battery can therefore act as a very powerful DC–DC converter. In addition, if the number of cells is continuously changed (on the input and/or output side) power conversion can also be AC/DC, AC/AC, or DC–AC with the frequency limited by that of the switching gear.

- Electric vehicles – Because flow batteries can be rapidly "recharged" by replacing the electrolyte, they can be used for applications where the vehicle needs to take on energy as fast as a combustion engined vehicle. A common problem found with most RFB chemistries in the EV applications is their low energy density which translated into a short driving range. Flow batteries based on highly soluble halates are a notable exception.

- Stand-alone power system – An example of this is in cellphone base stations where no grid power is available. The battery can be used alongside solar or wind power sources to compensate for their fluctuating power levels and alongside a generator to make the most efficient use of it to save fuel. Currently, flow batteries are being used in solar micro grid applications throughout the Caribbean.

# Rechargeable Battery

A rechargeable lithium polymer mobile phone battery

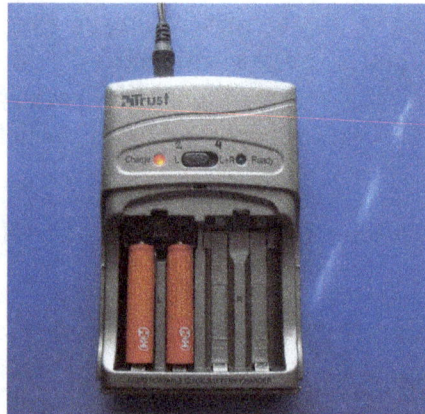

A common consumer battery charger for rechargeable AA and AAA batteries

A rechargeable battery, storage battery, secondary cell, or accumulator is a type of electrical battery which can be charged, discharged into a load, and recharged many times, while a non-rechargeable or primary battery is supplied fully charged, and discarded once discharged. It is composed of one or more electrochemical cells. The term "accumulator" is used as it accumulates and stores energy through a reversible electrochemical reaction. Rechargeable batteries are produced in many different shapes and sizes, ranging from button cells to megawatt systems connected to stabilize an electrical distribution network. Several different combinations of electrode materials and electrolytes are used, including lead–acid, nickel cadmium (NiCd), nickel metal hydride (NiMH), lithium ion (Li-ion), and lithium ion polymer (Li-ion polymer).

Rechargeable batteries typically initially cost more than disposable batteries, but have a much lower total cost of ownership and environmental impact, as they can be recharged inexpensively many times before they need replacing. Some rechargeable battery types are available in the same sizes and voltages as disposable types, and can be used interchangeably with them.

## Usage and Applications

Devices which use rechargeable batteries include automobile starters, portable consumer devices, light vehicles (such as motorized wheelchairs, golf carts, electric bicycles, and electric forklifts), tools, uninterruptible power supplies, and battery storage power stations. Emerging applications in hybrid internal combustion-battery and electric vehicles drive the technology to reduce cost, weight, and size, and increase lifetime.

Cylindrical cell (18650) prior to assembly.
Several thousand of them (lithium ion) form the Tesla Model S battery.

Lithium ion battery monitoring electronics (over- and discharge protection)

Older rechargeable batteries self-discharge relatively rapidly, and require charging before first use; some newer low self-discharge NiMH batteries hold their charge for many months, and are typically sold factory-charged to about 70% of their rated capacity.

Battery storage power stations use rechargeable batteries for load-leveling (storing electric energy at times of low demand for use during peak periods) and for renewable energy uses (such as storing power generated from photovoltaic arrays during the day to be used at night). Load-leveling reduces the maximum power which a plant must be able to generate, reducing capital cost and the need for peaking power plants.

The US National Electrical Manufacturers Association estimated in 2006 that US demand for rechargeable batteries was growing twice as fast as demand for disposables.

Small rechargeable batteries can power portable electronic devices, power tools, appliances, and so on. Heavy-duty batteries power electric vehicles, ranging from scooters to locomotives and ships. They are used in distributed electricity generation and in stand-alone power systems.

## Charging and Discharging

A solar-powered charger for rechargeable AA batteries

During charging, the positive active material is oxidized, producing electrons, and the negative material is reduced, consuming electrons. These electrons constitute the current flow in the external circuit. The electrolyte may serve as a simple buffer for internal ion flow between the electrodes, as in lithium-ion and nickel-cadmium cells, or it may be an active participant in the electrochemical reaction, as in lead–acid cells.

The energy used to charge rechargeable batteries usually comes from a battery charger using AC mains electricity, although some are equipped to use a vehicle's 12-volt DC power outlet. Regardless, to store energy in a secondary cell, it has to be connected to a DC voltage source. The negative terminal of the cell has to be connected to the negative terminal of the voltage source and the positive terminal of the voltage source with the positive terminal of the battery. Further, the voltage output of the source must be higher than that of the battery, but not *much* higher: the greater the difference between the power source and the battery's voltage capacity, the faster the charging process, but also the greater the risk of overcharging and damaging the battery.

Chargers take from a few minutes to several hours to charge a battery. Slow "dumb" chargers without voltage or temperature-sensing capabilities will charge at a low rate, typically taking 14 hours or more to reach a full charge. Rapid chargers can typically charge cells in two to five hours, depending on the model, with the fastest taking as little as fifteen minutes. Fast chargers must have multiple ways of detecting when a cell reaches full charge (change in terminal voltage, temperature, etc.) to stop charging before harmful overcharging or overheating occurs. The fastest chargers often incorporate cooling fans to keep the cells from overheating.

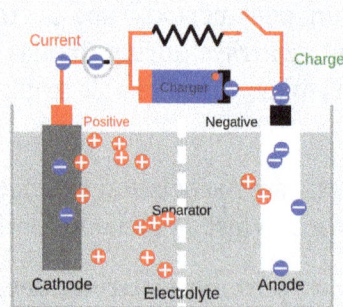

Diagram of the charging of a secondary cell or battery

Battery charging and discharging rates are often discussed by referencing a "C" rate of current. The C rate is that which would theoretically fully charge or discharge the battery in one hour. For example, trickle charging might be performed at C/20 (or a "20 hour" rate), while typical charging and discharging may occur at C/2 (two hours for full capacity). The available capacity of electrochemical cells varies depending on the discharge rate. Some energy is lost in the internal resistance of cell components (plates, electrolyte, interconnections), and the rate of discharge is limited by the speed at which chemicals in the cell can move about. For lead-acid cells, the relationship between time and discharge rate is described by Peukert's law; a lead-acid cell that can no longer sustain a usable terminal voltage at a high current may still have usable capacity, if discharged at a much lower rate. Data sheets for rechargeable cells often list the discharge capacity on 8-hour or 20-hour or other stated time; cells for uninterruptible power supply systems may be rated at 15 minute discharge.

Battery manufacturers' technical notes often refer to voltage per cell (VPC) for the individual cells that make up the battery. For example, to charge a 12 V lead-acid battery (containing 6 cells of 2 V each) at 2.3 VPC requires a voltage of 13.8 V across the battery's terminals.

Non-rechargeable alkaline and zinc–carbon cells output 1.5V when new, but this voltage drops with use. Most NiMH AA and AAA cells are rated at 1.2 V, but have a flatter discharge curve than alkalines and can usually be used in equipment designed to use alkaline batteries.

## Damage from Cell Reversal

Subjecting a discharged cell to a current in the direction which tends to discharge it further to the point the positive and negative terminals switch polarity causes a condition called *cell reversal*. Generally, pushing current through a discharged cell in this way causes undesirable and irreversible chemical reactions to occur, resulting in permanent damage to the cell. Cell reversal can occur under a number of circumstances, the two most common being:

- When a battery or cell is connected to a charging circuit the wrong way around.

- When a battery made of several cells connected in series is deeply discharged.

In the latter case, the problem occurs due to the different cells in a battery having slightly different capacities. When one cell reaches discharge level ahead of the rest, the remaining cells will force the current through the discharged cell.

Many battery-operated devices have a low-voltage cutoff that prevents deep discharges from occurring that might cause cell reversal.

Cell reversal can occur to a weakly charged cell even before it is fully discharged. If the battery drain current is high enough, the cell's internal resistance can create a resistive voltage drop that is greater than the cell's forward emf. This results in the reversal of the cell's polarity while the current is flowing. The higher the required discharge rate of a battery, the better matched the cells should be, both in the type of cell and state of charge, in order to reduce the chances of cell reversal.

In some situations, such as when correcting Ni-Cad batteries that have been previously overcharged, it may be desirable to fully discharge a battery. To avoid damage from the cell reversal effect, it is necessary to access each cell separately: each cell is individually discharged by connecting a load clip across the terminals of each cell, thereby avoiding cell reversal.

## Damage During Storage in Fully Discharged State

If a multi-cell battery is fully discharged, it will often be damaged due to the cell reversal effect mentioned above. It is possible however to fully discharge a battery without causing cell reversal—either by discharging each cell separately, or by allowing each cell's internal leakage to dissipate its charge over time.

Even if a cell is brought to a fully discharged state without reversal, however, damage may occur over time simply due to remaining in the discharged state. An example of this is the sulfation that occurs in lead-acid batteries that are left sitting on a shelf for long periods. For this reason it is often recommended to charge a battery that is intended to remain in storage, and to maintain its charge level by periodically recharging it. Since damage may also occur if the battery is over-charged, the optimal level of charge during storage is typically around 30% to 70%.

## Depth of Discharge

Depth of discharge (DOD) is normally stated as a percentage of the nominal ampere-hour capacity; 0% DOD means no discharge. As the usable capacity of a battery system depends on the rate of discharge and the allowable voltage at the end of discharge, the depth of discharge must be qualified to show the way it is to be measured. Due to variations during manufacture and aging, the DOD for complete discharge can change over time or number of charge cycles. Generally a rechargeable battery system will tolerate more charge/discharge cycles if the DOD is lower on each cycle.

## Lifespan and Cycle Stability

If batteries are used repeatedly even without mistreatment, they lose capacity as the number of charge cycles increases, until they are eventually considered to have reached the end of their useful life.

Lithium iron phosphate batteries reach according to the manufacturer more than 5000 cycles at respective depth of discharge of 70%. After 7500 cycles with discharge of 85% this still have a spare capacity of at least 80% at a rate of 1 C; which corresponds with a full cycle per day to a lifetime of min. 20.5 years.

The lithium iron phosphate battery Sony Fortelion has after 10,000 cycles at 100% discharge level still a residual capacity of 71%. This battery has been on the market since 2009.

Lithium-ion batteries have partly a very high cycle resistance of more than 10,000 charge and discharge cycles and a long service life of up to 20 years.

Plug in America has among drivers of the Tesla Roadster, a survey carried out with respect to the service life of the installed battery. It was found that after 100,000 miles = 160,000 km, the battery still had a remaining capacity of 80 to 85 percent. This was regardless of in which climate zone the car is moved. The Tesla Roadster was built and sold between 2008 and 2012. For its 85-kWh batteries in the Tesla Model S Tesla are 8-year warranty with unlimited mileage.

Varta Storage guarantees its engion battery systems for 14,000 full cycles and a service life of 10 years.

As of 2017, the best-selling electric car is the Nissan Leaf, which is produced since of 2010. Nissan stated in 2015 that until then only 0.01 percent of batteries had to be replaced because of failures or problems and then only because of externally inflicted damage. There are few vehicles that have already covered more than 200,000 km away. These have no problems with the battery.

## Recharging Time

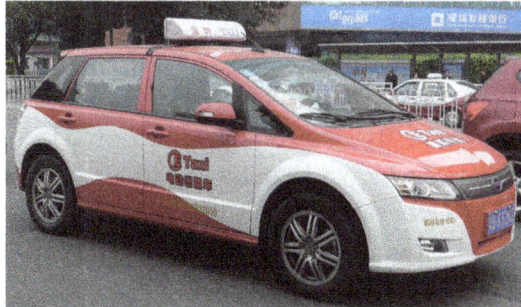

BYD e6 taxi. Recharging in 15 Minutes to 80 percent

Electric cars like Tesla Model S, Renault Zoe, BMW i3, etc. can recharge their batteries at quick charging stations within 30 minutes to 80 percent. The Porsche Mission E will be able to charge to 80 percent within 15 minutes.

In laboratories the company StoreDot from Israel reportedly demonstrated the first lab samples of unspecified batteries that can, as of April 2014, be charged in 30 seconds in mobile phones.

Researchers from Singapore in 2014 developed a battery that can be recharged in 2 minutes to 70 percent. The batteries rely on lithium-ion technology. However, the anode and the negative pole in the battery is no longer made of graphite, but a titanium dioxide gel. The gel accelerates the chemical reaction significantly, thus ensuring a faster charging. In particular, these batteries are to be used in electric cars. Already in 2012 researchers at the Ludwig-Maximilian-University in Munich have discovered the basic principle.

Scientists at Stanford University in California have developed a battery that can be charged within one minute. The anode is made of aluminium and the cathode made of graphite.

The electric car Volar-e of the company Applus + IDIADA, based on the Rimac Concept One, contains lithium iron phosphate batteries that can be recharged in 15 minutes.

According to the manufacturer BYD the lithium iron phosphate battery of the electric car e6 is charged at a fast charging station within 15 minutes to 80%, after 40 minutes at 100%.

In 2005, handheld device battery designs by Toshiba were claimed to be able to accept an 80% charge in as little as 60 seconds.

Scientists of university of Oslo from Norway have developed a battery which can be recharged less than one second. According to the scientists this battery would be interesting for example for city buses, which could be loaded at each bus stop, and thus would require only a relatively small battery. A disadvantage is, according to the researchers that the bigger the battery, the greater must be the charging current. Thus, the battery can not be very big. According to the researchers of the new battery could also be used as a buffer in sports car to provide power in the short term. For now, however, the researchers think of applications in small and micro devices.

According to the manufacturer battery of the smartphone OnePlus 3 can be charged from 0 to 60 percent within 30 minutes.

## Price History

Lead-acid batteries typically cost €100 / kWh. Li-Ion batteries cost in January 2014, however, typically around €110 / kWh (150 USD / kWh). The prices for Li-Ion batteries have, since 2011, dropped significantly (2011: €500 / kWh, 2012: €350 / kWh, 2013: €200 / kWh) At a conference for electric mobility October 2013 mentioned the trend researcher Lars Thomsen, that Tesla has built its battery at the time 200 USD / kWh (equivalent to €148 / kWh). for the planned for autumn 2016 e-mobile Bolt expects General Motors 145 USD / kWh, and a reduction to 100 USD / kWh by 2022. Reasons for the price decline is the increasing mass production, which has reduced costs through better technologies and economies of scale.

In the German retail LiFePO4 battery cells (as of January 2015) are available from about 420 € / kWh (1.35 € / Ah).

The Powerpack of Tesla costs in spring 2016 250 USD per kWh.

According to a study by McKinsey, the rate of rechargeable battery fell by 80 percent between 2010 and 2016.

## Active Components

The active components in a secondary cell are the chemicals that make up the positive and negative active materials, and the electrolyte. The positive and negative are made up of different materials, with the positive exhibiting a reduction potential and the negative having an oxidation potential. The sum of these potentials is the standard cell potential or voltage.

In primary cells the positive and negative electrodes are known as the cathode and anode, respectively. Although this convention is sometimes carried through to rechargeable systems—especially with lithium-ion cells, because of their origins in primary lithium cells—this practice can lead to confusion. In rechargeable cells the positive electrode is the cathode on discharge and the anode on charge, and vice versa for the negative electrode.

## Types

The lead–acid battery, invented in 1859 by French physicist Gaston Planté, is the oldest type of rechargeable battery. Despite having a very low energy-to-weight ratio and a low energy-to-volume ratio, its ability to supply high surge currents means that the cells have a relatively large power-to-weight ratio. These features, along with the low cost, makes it attractive for use in motor vehicles to provide the high current required by automobile starter motors.

The nickel–cadmium battery (NiCd) was invented by Waldemar Jungner of Sweden in 1899. It uses nickel oxide hydroxide and metallic cadmium as electrodes. Cadmium is a toxic element, and was banned for most uses by the European Union in 2004. Nickel–cadmium batteries have been almost completely superseded by nickel–metal hydride (NiMH) batteries.

The nickel–metal hydride battery (NiMH) became available in 1989. These are now a common consumer and industrial type. The battery has a hydrogen-absorbing alloy for the negative electrode instead of cadmium.

The lithium-ion battery was introduced in the market in 1991, and it is the choice in most consumer electronics and has the best energy density and a very slow loss of charge when not in use. It does have drawbacks too, particularly the risk of unexpected ignition from the heat generated by the battery. Such incidents are rare and according to experts, they can be minimized "via appropriate design, installation, procedures and layers of safeguards" so the risk is acceptable.

Lithium-ion polymer batteries (LiPo) are light in weight, offer slightly higher energy density than Li-ion at slightly higher cost, and can be made in any shape. They are available but have not displaced Li-ion in the market. A primary use is for LiPo batteries is in powering remote-controlled cars, boats and airplanes. LiPo packs are readily available on the consumer market, in various configurations, up to 44.4v, for powering certain certain R/C vehicles and helicopters or drones. Some test reports warn of the risk of fire when the batteries are not used in accordance with the instructions. Independent reviews of the technology discuss the risk of fire and explosion from Lithium-ion batteries under certain conditions because they use liquid electrolytes.

## Solid State

On 28 February 2017, The University of Texas at Austin issued a press release about a new type of solid-state battery, developed by a team of engineers led by Lithium-ion (Li-Ion) inventor John Goodenough, "that could lead to safer, faster-charging, longer-lasting rechargeable batteries for handheld mobile devices, electric cars and stationary energy storage". More specifics about the new technology were published by Goodenough and his team of engineers on 9 December 2016 in the peer-reviewed scientific journal Energy & Environmental Science.

Independent reviews of the technology discuss the risk of fire and explosion from Lithium-ion batteries under certain conditions because they use liquid electrolytes. The newly developed battery should be safer since it uses glass electrolytes, that should eliminate short circuits. (More specifically, the battery uses glass electrolytes that enable the use of an alkali-metal anode without the formation of dendrites.)

The solid-state battery is also said to have "three times the energy density" increasing its useful life in electric vehicles, for example. It should also be more ecologically sound since the technology uses less expensive, earth-friendly materials such as sodium extracted from seawater. Another claimed benefit is longer useable life; ("the cells have demonstrated more than 1,200 cycles with low cell resistance"). The research and prototypes are not expected to lead to a commercially viable product in the near future, if ever, according to Chris Robinson of LUX Research. "This will have no tangible effect on electric vehicle adoption in the next 15 years, if it does at all. A key hurdle that many solid-state electrolytes face is lack of a scalable and cost-effective manufacturing process," he told The American Energy News in an e-mail.

## Other Experimental Types

| Type | Voltage[a] | Energy density[b] | | | Power[c] | E/$[e] | Self-disch.[f] | Charge Efficiency | Cycles[g] | Life[h] |
|---|---|---|---|---|---|---|---|---|---|---|
| | (V) | (MJ/kg) | (Wh/kg) | (Wh/L) | (W/kg) | (Wh/$) | (%/month) | (%) | (#) | (years) |
| Lithium sulfur | 2.0 | 0.94-1.44 | 400 | 350 | | | | | ~1400 | |
| Sodium-ion | 3.6 | | | 30 | | 3.3 | | | 5000+ | Testing |
| Thin film lithium | ? | | 300 | 959 | 6000 | ?[p] | | | 40000 | |
| Zinc-bromide | | 0.27-0.31 | 75-85 | | | | | | | |
| Zinc-cerium | 2.5 | | | | | | | | | Under testing |
| Vanadium redox | 1.15-1.55 | 0.09-0.13 | 25-35 | | | | 20% | | 14,000 | 10 (stationary) |
| Sodium-sulfur | | 0.54 | 150 | | | | | 89–92% | 2500—4500 | |
| Molten salt | 2.58 | 0.25-1.04 | 70-290 | 160 | 150-220 | 4.54 | | | 3000+ | <=20 |
| Silver-zinc | 1.86 | 0.47 | 130 | 240 | | | | | | |
| Quantum Battery (oxide semiconductor) | 1.5-3 | | | 500 | 8000(W/L) | | | | 100,000 | |

‡ citations are needed for these parameters

Notes

    [a] Nominal cell voltage in V.

    [b] Energy density = energy/weight or energy/size, given in three different units

    [c] Specific power = power/weight in W/kg

    [e] Energy/consumer price in W·h/US$ (approximately)

    [f] Self-discharge rate in %/month

    [g] Cycle durability in number of cycles

    [h] Time durability in years

    [i] VRLA or recombinant includes gel batteries and absorbed glass mats

    [p] Pilot production

The lithium–sulfur battery was developed by Sion Power in 1994. The company claims superior energy density to other lithium technologies.

The thin film battery (TFB) is a refinement of lithium ion technology by Excellatron. The developers claim a large increase in recharge cycles to around 40,000 and higher charge and discharge rates, at least 5 $C$ charge rate. Sustained 60 $C$ discharge and 1000$C$ peak discharge rate and a significant increase in specific energy, and energy density.

A smart battery has voltage monitoring circuit built inside. Carbon foam-based lead acid battery: Firefly Energy developed a carbon foam-based lead acid battery with a reported energy density of 30-40% more than their original 38 Wh/kg, with long life and very high power density.

UltraBattery, a hybrid lead-acid battery and ultracapacitor invented by Australia's national science organisation CSIRO, exhibits tens of thousands of partial state of charge cycles and has outperformed traditional lead-acid, lithium and NiMH-based cells when compared in testing in this mode against variability management power profiles. UltraBattery has kW and MW-scale installations in place in Australia, Japan and the U.S.A. It has also been subjected to extensive testing in hybrid electric vehicles and has been shown to last more than 100,000 vehicle miles in on-road commercial testing in a courier vehicle. The technology is claimed to have a lifetime of 7 to 10 times that of conventional lead-acid batteries in high rate partial state-of-charge use, with safety and environmental benefits claimed over competitors like lithium-ion. Its manufacturer suggests an almost 100% recycling rate is already in place for the product.

The potassium-ion battery delivers around a million cycles, due to the extraordinary electrochemical stability of potassium insertion/extraction materials such as Prussian blue.

The sodium-ion battery is meant for stationary storage and competes with lead–acid batteries. It aims at a low total cost of ownership per kWh of storage. This is achieved by a long and stable lifetime. The effective number of cycles is above 5000 and the battery is not damaged by deep discharge. The energy density is rather low, somewhat lower than lead–acid.

The quantum battery (oxide semiconductor) was developed by MJC. It is a small, lightweight cell with a multi-layer film structure and high energy and high power density. It is incombustible, has no electrolyte and generates a low amount of heat during charge. Its unique feature is its ability to capture electrons physically rather than chemically.

In 2007, Yi Cui and colleagues at Stanford University's Department of Materials Science and Engineering discovered that using silicon nanowires as the anode of a lithium-ion battery increases the anode's volumetric charge density by up to a factor of 10, leading to the development of the nanowire battery.

Another development is the paper-thin flexible self-rechargeable battery combining a thin-film organic solar cell with an extremely thin and highly flexible lithium-polymer battery, which recharges itself when exposed to light.

Ceramatec, a research and development unit of CoorsTek, as of 2009 was testing a battery comprising a chunk of solid sodium metal mated to a sulfur compound by a paper-thin ceramic membrane which conducts ions back and forth to generate a current. The company claimed that it could fit about 40 kilowatt hours of energy into a package about the size of a refrigerator, and operate below 90 °C; and that their battery would allow about 3,650 discharge/recharge cycles (or roughly 1 per day for one decade).

Battery electrodes can be microscopically viewed while bathed in wet electrolytes, resembling conditions inside operating batteries.

In 2014, an Israeli company, StoreDot, claimed to be able to charge batteries in 30 seconds.

Secondary magnesium battery types are an active (2015) topic of research, as a replacement for lithium ion cells.

Aluminium-ion battery types had big success in 2015 in research.

## Alternatives

A rechargeable battery is only one of several types of rechargeable energy storage systems. Several alternatives to rechargeable batteries exist or are under development. For uses such as portable radios, rechargeable batteries may be replaced by clockwork mechanisms which are wound up by hand, driving dynamos, although this system may be used to charge a battery rather than to operate the radio directly. Flashlights may be driven by a dynamo directly. For transportation, uninterruptible power supply systems and laboratories, flywheel energy storage systems store energy in a spinning rotor for conversion to electric power when needed; such systems may be used to provide large pulses of power that would otherwise be objectionable on a common electrical grid.

Ultracapacitors—capacitors of extremely high value— are also used; an electric screwdriver which charges in 90 seconds and will drive about half as many screws as a device using a rechargeable battery was introduced in 2007, and similar flashlights have been produced. In keeping with the concept of ultracapacitors, betavoltaic batteries may be utilized as a method of providing a trickle-charge to a secondary battery, greatly extending the life and energy capacity of the battery system being employed; this type of arrangement is often referred to as a "hybrid betavoltaic power source" by those in the industry.

Ultracapacitors are being developed for transportation, using a large capacitor to store energy instead of the rechargeable battery banks used in hybrid vehicles. One drawback of capacitors compared to batteries is that the terminal voltage drops rapidly; a capacitor that has 25% of its initial energy left in it will have one-half of its initial voltage. By contrast, battery systems tend to have a terminal voltage that does not decline rapidly until nearly exhausted. The undesirable characteristic complicates the design of power electronics for use with ultracapacitors. However, there are potential benefits in cycle efficiency, lifetime, and weight compared with rechargeable systems. China started using ultracapacitors on two commercial bus routes in 2006; one of them is route 11 in Shanghai.

Flow batteries, used for specialized applications, are recharged by replacing the electrolyte liquid. A flow battery can be considered to be a type of rechargeable fuel cell.

## Research

Lithium Ion batteries normally have an anode made of graphite. Using an anode made of silicon (Si) can increase the capacity up to 6 times, because the Si-anode can accept much more Lithium-ion than a graphite-anode. A problem was, that the Si-anode expands 300–400% when charged. The Si-anode had only a small lifespan. Researchers of the university of Stuttgart (institute of photovoltaic (IPV), Prof. Dr. Jürgen H. Werner and his team) found a way making the Si-anode porous, so that accepting so many Lithium-ion will not longer increase the volume of the

Si-anode, so that the lifespan of the battery with Si-anode is now four times higher than batteries with graphite-anode. The battery is ready for production.

John B. Goodenough, one of the inventor of lithium-ion battery, recently (2017) helped develop the glass battery, a developmental battery with a glass electrolyte that is reported to exceed current lithium-ion batteries in energy density, operating temperature range, and safety.

## References

- Shiokawa, Y.; Yamana, H.; Moriyama, H. (2000). "An Application of Actinide Elements for a Redox Flow Battery". Journal of Nuclear Science and Technology. 37 (3): 253–256. doi:10.1080/18811248.2000.9714891

- Hayano, Ryugo S. (29 September 2009). "Development of a charged-particle accumulator using an RF confinement method" (report). Defense Technical Information Center. University of Tokyo. Retrieved 12 April 2015

- Alotto, P.; Guarnieri, M.; Moro, F. (2014). "Redox Flow Batteries for the storage of renewable energy: a review". Renewable & Sustainable Energy Reviews. 29: 325–335. doi:10.1016/j.rser.2013.08.001

- David Linden, Thomas B. Reddy (ed). Handbook Of Batteries 3rd Edition. McGraw-Hill, New York, 2002 ISBN 0-07-135978-8 chapter 22

- Leung, P. K.; Ponce-De-León, C.; Low, C. T. J.; Shah, A. A.; Walsh, F. C. (2011). "Characterization of a zinc–cerium flow battery". Journal of Power Sources. 196 (11): 5174–5185. doi:10.1016/j.jpowsour.2011.01.095

- "Chemists present an innovative redox-flow battery based on organic polymers and water". phys.org. Phys.org. 21 October 2015. Retrieved 6 December 2015

- Braga, M.H.; Grundish, N.S.; Murchison, A.J.; Goodenough, J.B. (9 December 2016). "Alternative strategy for a safe rechargeable battery". Energy and Environmental Science. doi:10.1039/C6EE02888H. Retrieved 15 March 2017

- Qi, Zhaoxiang; Koenig Jr., Gary M. (2016-08-15). "A carbon-free lithium-ion solid dispersion redox couple with low viscosity for redox flow batteries". Journal of Power Sources. 323: 97–106. doi:10.1016/j.jpowsour.2016.05.033

- Kevin Bullis (24 April 2014). "Nanoparticle Networks Promise Cheaper Batteries for Storing Renewable Energy". MIT Technology Review. Retrieved 24 September 2014

- Eftekhari, A.; Jian, Z.; Ji, X. (2017). "Potassium Secondary Batteries". ACS Applied Materials & Interfaces. 9: 4404–4419. doi:10.1021/acsami.6b07989

- Katerina E. Aifantis et al, High Energy Density Lithium Batteries: Materials, Engineering, Applications Wiley-VCH, 2010 ISBN 3-527-32407-0 page 66

- Tirone, Johnathan (15 March 2017). "Google's Schmidt Flags Promise in New Goodenough Battery". Bloomberg. Retrieved 21 March 2017

# An Overview of Grid Energy Storage

Grid energy storage seeks to store energy in an electrical grid. Compressed air energy storage and cryogenic energy storage are some of the ways to store energy in a grid. The aspects elucidated in this chapter are of vital importance, and provide a better understanding of energy storage.

## Grid Energy Storage

Grid energy storage (also called large-scale energy storage) is a collection of methods used to store electrical energy on a large scale within an electrical power grid. Electrical energy is stored during times when production (especially from intermittent power plants such as renewable electricity sources such as wind power, tidal power, solar power) exceeds consumption, and returned to the grid when production falls below consumption.

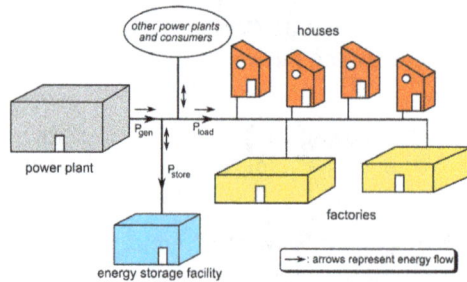

Simplified electrical grid with energy storage

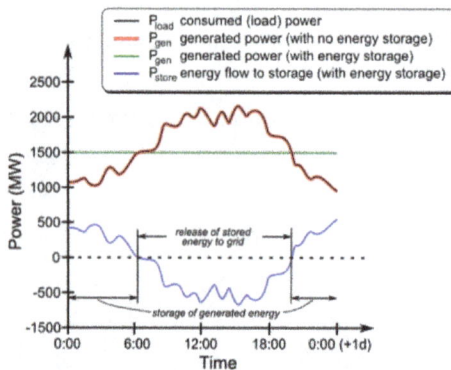

Simplified grid energy flow with and without idealized energy storage for the course of one day

As of 2017, by far the largest form of grid energy storage on grids is dammed hydroelectricity, with both conventional hydroelectric generation as well as pumped storage.

An alternative to grid storage is the use of peaking power plants to fill in demand gaps.

## Benefits of Storage and Managing Peak Load

The stores are used – feeding power to the grids – at times when consumption that cannot be deferred or delayed exceeds production. In this way, electricity production need not be drastically scaled up and down to meet momentary consumption – instead, transmission from the combination of generators plus storage facilities is maintained at a more constant level.

An alternate and complementary approach to achieve the similar effect as grid energy storage is to use a smart grid communication infrastructure to enable Demand response. These technologies shift electricity consumption and electricity production from one time (when it's not useful) to another (when it's in demand).

Any electrical power grid must match electricity production to consumption, both of which vary drastically over time. Any combination of energy storage and demand response has these advantages:

- fuel-based power plants (i.e. coal, oil, gas, nuclear) can be more efficiently and easily operated at constant production levels

- electricity generated by intermittent sources can be stored and used later, whereas it would otherwise have to be transmitted for sale elsewhere, or shut down

- peak generating or transmission capacity can be reduced by the total potential of all storage plus deferrable loads, saving the expense of this capacity

- more stable pricing – the cost of the storage and/or demand management is included in pricing so there is less variation in power rates charged to customers, or alternatively (if rates are kept stable by law) less loss to the utility from expensive on-peak wholesale power rates when peak demand must be met by imported wholesale power

- emergency preparedness – vital needs can be met reliably even with no transmission or generation going on while non-essential needs are deferred

Energy derived from solar, tidal and wind sources inherently varies – the amount of electricity produced varies with time of day, moon phase, season, and random factors such as the weather. Thus, renewables in the absence of storage present special challenges to electric utilities. While hooking up many separate wind sources can reduce the overall variability, solar is reliably not available at night, and tidal power shifts with the moon, so slack tides occur four times a day.

How much this affects any given utility varies significantly. In a summer peak utility, more solar can generally be absorbed and matched to demand. In winter peak utilities, to a lesser degree, wind correlates to heating demand and can be used to meet that demand. Depending on these factors, beyond about 20–40% of total generation, grid-connected intermittent sources such as solar power and wind turbines tend to require investment in grid interconnections, grid energy storage or demand side management.

In an electrical grid without energy storage, generation that relies on energy stored within fuels (coal, biomass, natural gas, nuclear) must be scaled up and down to match the rise and fall of electrical production from intermittent sources. While hydroelectric and natural gas plants can be quickly scaled up or down to follow the wind, coal and nuclear plants take considerable time to respond to load. Utilities with less natural gas or hydroelectric generation are thus more reliant on demand management, grid interconnections or costly pumped storage.

The French consulting firm Yole Développement estimates the "stationary storage" market could be a $13.5 billion opportunity by 2023, compared with less than $1 billion in 2015.

## Demand Side Management and Grid Storage

The demand side can also store electricity from the grid, for example charging a battery electric vehicle stores energy for a vehicle and storage heaters, district heating storage or ice storage provide thermal storage for buildings. At present this storage serves only to shift consumption to the off-peak time of day, no electricity is returned to the grid.

The need for grid storage to provide peak power is reduced by demand side time of use pricing, one of the benefits of smart meters. At the household level, consumers may choose less expensive off-peak times for clothes washer/dryers, dishwashers, showers and cooking. As well commercial and industrial users will take advantage of cost savings by deferring some processes to off-peak times.

Regional impacts from the unpredictable operation of wind power has created a new need for interactive demand response, where the utility communicates with the demand. Historically this was only done in cooperation with large industrial consumers, but now may be expanded to entire grids. For instance a few large scale projects in Europe link variations in wind power to change industrial food freezer loads, causing small variations in temperature. If communicated on a grid-wide scale, small changes to heating/cooling temperatures would instantly change consumption across the grid.

A report released in December 2013 by the United States Department of Energy further describes the potential benefits of energy storage and demand side technologies to the electric grid: "Modernizing the electric system will help the nation meet the challenge of handling projected energy needs—including addressing climate change by integrating more energy from renewable sources and enhancing efficiency from non-renewable energy processes. Advances to the electric grid must maintain a robust and resilient electricity delivery system, and energy storage can play a significant role in meeting these challenges by improving the operating capabilities of the grid, lowering cost and ensuring high reliability, as well as deferring and reducing infrastructure investments. Finally, energy storage can be instrumental for emergency preparedness because of its ability to provide backup power as well as grid stabilization services." The report was written by a core group of developers representing Office of Electricity Delivery and Energy Reliability, ARPA-E, Office of Science, Office of Energy Efficiency and Renewable Energy, Sandia National Laboratories, and Pacific Northwest National Laboratory; all of whom are engaged in the development of grid energy storage.

## Forms

## Compressed Air

Another grid energy storage method is to use off-peak or renewably generated electricity to compress air, which is usually stored in an old mine or some other kind of geological feature. When electricity demand is high, the compressed air is heated with a small amount of natural gas and then goes through turboexpanders to generate electricity.

Compressed air storage is typically around 60–90% efficient

## Liquid Air

Another electricity storage method is to compress and cool air, turning it into liquid air, which can be stored, and expanded when needed, turning a turbine, generating electricity, with a storage efficiency of up to 70%.

## Batteries

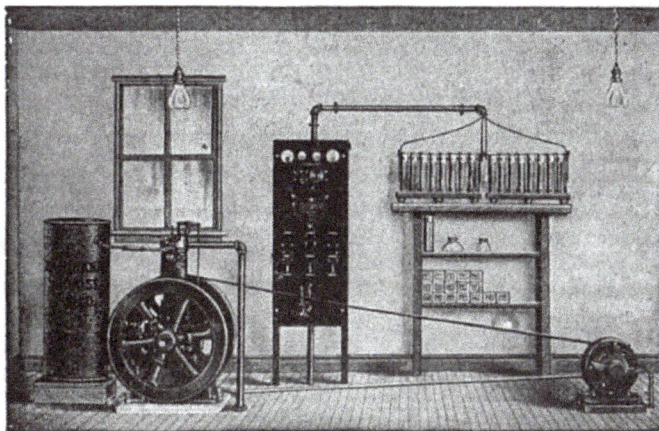

A 900 watt direct current light plant using 16 separate lead acid battery cells (32 volts)

Battery storage was used in the early days of direct current electric power. Where AC grid power was not readily available, isolated lighting plants run by wind turbines or internal combustion engines provided lighting and power to small motors. The battery system could be used to run the load without starting the engine or when the wind was calm. A bank of lead-acid batteries in glass jars both supplied power to illuminate lamps, as well as to start an engine to recharge the batteries. Battery storage technology is typically around 70–>85% efficient.

Battery systems connected to large solid-state converters have been used to stabilize power distribution networks. Some grid batteries are co-located with renewable energy plants, either to smooth the power supplied by the intermittent wind or solar output, or to shift the power output into other hours of the day when the renewable plant cannot produce power directly. These hybrid systems (generation + storage) can either alleviate the pressure on the grid when connecting renewable of be used to reach self-sufficiency and work "off-the-grid" .

Contrary to electric vehicle applications, batteries for stationary storage do not suffer from mass or volume constraints. However, due to the large amounts of energy and power implied, the cost per power or energy unit is crucial. The relevant metrics to assess the interest of a technology for grid-scale storage is the $/Wh (or $/W) rather than the Wh/kg (or W/kg). The electrochemical grid storage was made possible thanks to the development of the electric vehicle, that induced a fast decrease in the production costs of batteries below $300/kWh. By optimizing the production chain, major industrials aim to reach $150/kWh by the end of 2020. These batteries rely on a

Li-Ion technology, which is suited for mobile applications (high cost, high density). Technologies optimized for the grid should focus on low cost and low density.

## Grid-oriented Battery Technologies

Sodium-Ion batteries are a cheap and sustainable alternative to Li-ion, because sodium is far more abundant and cheap than lithium, but it has a lower power density. However, they are still on the early stages of their development.

Automotive-oriented technologies rely on solid electrodes, which feature a high energy density but require an expensive manufacturing process. Liquid electrodes represent a cheaper and less dense alternative as they do not need any processing.

## Molten-state Batteries

These batteries are composed of two molten metal alloys separated by an electrolyte. They are simple to manufacture but require a temperature of several hundred degree Celsius to keep the alloy in a liquid state. This technology includes ZEBRA, Sodium-sulfur batteries and liquid metal. Sodium sulphur batteries are being used for grid storage in Japan and in the United States. The electrolyte is composed of solid beta alumina. The liquid metal battery, developed by the group of Pr. Sadoway, uses molten alloys of Magnesium and antimony separated by an electrically insulating molten salt. It is still is the prototyping phase.

## Flow Batteries

In rechargeable flow batteries, the liquid electrodes are composed of transition metals in water at room temperature. They can be used as a rapid-response storage medium. Vanadium redox batteries is another flow battery. They are installed at Huxley Hill wind farm (Australia), Tomari Wind Hills at Hokkaidō (Japan), as well as in non-wind farm applications. A 12 MW·h flow battery was to be installed at the Sorne Hill wind farm (Ireland). These storage systems are designed to smooth out transient wind fluctuations. Hydrogen Bromide has been proposed for use in a utility-scale flow-type battery.

## Examples

For example, in Puerto Rico a system with a capacity of 20 megawatts for 15 minutes (5 megawatt hour) stabilizes the frequency of electric power produced on the island. A 27 megawatt 15-minute (6.75 megawatt hour) nickel-cadmium battery bank was installed at Fairbanks Alaska in 2003 to stabilize voltage at the end of a long transmission line.

In 2016 a zinc-ion battery was proposed for use in grid storage applications.

In 2017 the California Public Utilities Commission installed 396 refrigerator-sized stacks of Tesla batteries at the Mira Loma substation in Ontario, California. The stacks are deployed in two modules of 10MW each (20MW in total), each capable of running for 4 hours, thus adding up to 80MWh of storage. The array is capable of powering 15,000 homes for over four hours.

BYD proposes to use conventional consumer battery technologies such as lithium iron phosphate

(LiFePO4) battery, connecting many batteries in parallel.

The largest grid storage batteries in the United States include the 31.5MW battery at Grand Ridge Power plant in Illinois and the 31.5 MW battery at Beech Ridge, West Virginia. Two batteries under construction in 2015 include the 400MWh (100MW for 4 hours) Southern California Edison project and the 52 MWh project on Kauai, Hawaii to entirely time shift a 13MW solar farm's output to the evening. Two batteries are in Fairbanks, Alaska (40 MW for 7 minutes using Ni-Cd cells), and in Notrees, Texas (36 MW for 40 minutes using lead-acid batteries). A 13 MWh battery made of used batteries from Daimler's Smart electric drive cars is being constructed in Lünen, Germany, with an expected second life of 10 years.

In 2015, a 221 MW battery storage was installed in the USA, with total capacity expected to reach 1.7 GW in 2020.

| Technology Comparison for Grid-Level Applications | | | | | | |
|---|---|---|---|---|---|---|
| Technology | Moving Parts | Operation at Room Temperature | Flammable | Toxic Materials | In production | Rare metals |
| Vanadium flow | Yes | Yes | No | Yes | Yes | No |
| Liquid Metal | No | No | Yes | No | No | No |
| Sodium-Ion | No | No | Yes | No | No | No |
| Lead-Acid | No | Yes | No | Yes | Yes | No |
| Sodium-sulfur batteries | No | No | No | Yes | Yes | No |
| Ni-Cd | No | Yes | No | Yes | Yes | Yes |
| Al-ion | No | Yes | No | No | No | No |
| Li-ion | No | Yes | Yes | No | Yes | No |

## Electric Vehicles

Nissan Leaf, the world's top-selling highway-capable electric car as of 2015

Companies are researching the possible use of electric vehicles to meet peak demand. A parked and plugged-in electric vehicle could sell the electricity from the battery during peak loads and charge either during night (at home) or during off-peak.

Plug-in hybrid or electric cars could be used for their energy storage capabilities. Vehicle-to-grid technology can be employed, turning each vehicle with its 20 to 50 kWh battery pack into a distributed load-balancing device or emergency power source. This represents 2 to 5 days per vehicle of average household requirements of 10 kWh per day, assuming annual consumption of 3650 kWh. This quantity of energy is equivalent to between 40 and 300 miles (64 and 483 km) of range in such vehicles consuming 0.5 to 0.16 kWh per mile. These figures can be achieved even in home-made electric vehicle conversions. Some electric utilities plan to use old plug-in vehicle batteries (sometimes resulting in a giant battery) to store electricity However, a large disadvantage of using vehicle to grid energy storage is the fact that each storage cycle stresses the battery with one complete charge-discharge cycle. Conventional (cobalt-based) lithium ion batteries break down with the number of cycles – newer li-ion batteries do not break down significantly with each cycle, and so have much longer lives. One approach is to reuse unreliable vehicle batteries in dedicated grid storage as they are expected to be good in this role for ten years. If such storage is done on a large scale it becomes much easier to guarantee replacement of a vehicle battery degraded in mobile use, as the old battery has value and immediate use.

## Flywheel

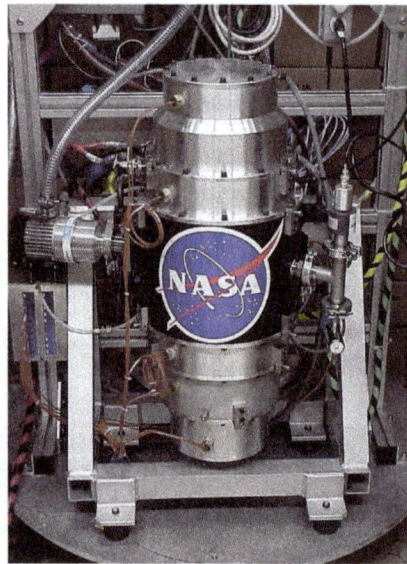

NASA G2 flywheel

Mechanical inertia is the basis of this storage method. When the electric power flows into the device, an electric motor accelerates a heavy rotating disc. The motor acts as a generator when the flow of power is reversed, slowing down the disc and producing electricity. Electricity is stored as the kinetic energy of the disc. Friction must be kept to a minimum to prolong the storage time. This is often achieved by placing the flywheel in a vacuum and using magnetic bearings, tending to make the method expensive. Greater flywheel speeds allow greater storage capacity but require strong materials such as steel or composite materials to resist the centrifugal forces. The ranges of power and energy storage technology that make this method economic, however, tends to make flywheels unsuitable for general power system application; they are probably best suited to load-leveling applications on railway power systems and for improving power quality in renewable energy systems such as the 20MW system in Ireland.

Applications that use flywheel storage are those that require very high bursts of power for very short durations such as tokamak and laser experiments where a motor generator is spun up to operating speed and is partially slowed down during discharge.

Flywheel storage is also currently used in the form of the Diesel rotary uninterruptible power supply to provide uninterruptible power supply systems (such as those in large datacenters) for ride-through power necessary during transfer – that is, the relatively brief amount of time between a loss of power to the mains and the warm-up of an alternate source, such as a diesel generator.

This potential solution has been implemented by EDA in the Azores on the islands of Graciosa and Flores. This system uses an 18 megawatt-second flywheel to improve power quality and thus allow increased renewable energy usage. As the description suggests, these systems are again designed to smooth out transient fluctuations in supply, and could never be used to cope with an outage exceeding a couple of days.

Powercorp in Australia have been developing applications using wind turbines, flywheels and low load diesel (LLD) technology to maximise the wind input to small grids. A system installed in Coral Bay, Western Australia, uses wind turbines coupled with a flywheel based control system and LLDs to achieve better than 60% wind contribution to the town grid.

## Hydrogen

Hydrogen is being developed as an electrical energy storage medium. Hydrogen is produced, then compressed or liquefied, cryogenicly stored at −252.882 °C, and then converted back to electrical energy or heat. Hydrogen can be used as a fuel for portable (vehicles) or stationary energy generation. Compared to pumped water storage and batteries, hydrogen has the advantage that it is a high energy density fuel.

Hydrogen can be produced either by reforming natural gas with steam or by the electrolysis of water into hydrogen and oxygen. Reforming natural gas produces carbon dioxide as a by-product. High temperature electrolysis and high pressure electrolysis are two techniques by which the efficiency of hydrogen production may be able to be increased. Hydrogen is then converted back to electricity in an internal combustion engine, or a fuel cell.

The AC-to-AC efficiency of hydrogen storage has been shown to be on the order of 20 to 45%, which imposes economic constraints. The price ratio between purchase and sale of electricity must be at least proportional to the efficiency in order for the system to be economic. Hydrogen fuel cells can respond quickly enough to correct rapid fluctuations in electricity demand or supply and regulate frequency. Whether hydrogen can use natural gas infrastructure depends on the network construction materials, standards in joints, and storage pressure.

The equipment necessary for hydrogen energy storage includes an electrolysis plant, hydrogen compressors or liquifiers, and storage tanks.

Biohydrogen is a process being investigated for producing hydrogen using biomass.

Micro combined heat and power (microCHP) can use hydrogen as a fuel.

Some nuclear power plants may be able to benefit from a symbiosis with hydrogen production. High temperature (950 to 1,000 °C) gas cooled nuclear generation IV reactors have the potential to electrolyze hydrogen from water by thermochemical means using nuclear heat as in the sulfur-iodine cycle. The first commercial reactors are expected in 2030.

A community based pilot program using wind turbines and hydrogen generators was started in 2007 in the remote community of Ramea, Newfoundland and Labrador. A similar project has been going on since 2004 in Utsira, a small Norwegian island municipality.

## Underground Hydrogen Storage

Underground hydrogen storage is the practice of hydrogen storage in underground caverns, salt domes and depleted oil and gas fields. Large quantities of gaseous hydrogen have been stored in underground caverns by Imperial Chemical Industries (ICI) for many years without any difficulties. The European project Hyunder indicated in 2013 that for the storage of wind and solar energy an additional 85 caverns are required as it can't be covered by PHES and CAES systems.

## Power to Gas

Power to gas is a technology which converts electrical power to a gas fuel. There are 2 methods, the first is to use the electricity for water splitting and inject the resulting hydrogen into the natural gas grid. The second less efficient method is used to convert carbon dioxide and water to methane, using electrolysis and the Sabatier reaction. The excess power or off peak power generated by wind generators or solar arrays is then used for load balancing in the energy grid. Using the existing natural gas system for hydrogen, fuel cell maker Hydrogenics and natural gas distributor Enbridge have teamed up to develop such a power to gas system in Canada.

Pipeline storage of hydrogen where a natural gas network is used for the storage of hydrogen. Before switching to natural gas, the German gas networks were operated using towngas, which for the most part consisted of hydrogen. The storage capacity of the German natural gas network is more than 200,000 GW·h which is enough for several months of energy requirement. By comparison, the capacity of all German pumped storage power plants amounts to only about 40 GW·h. The transport of energy through a gas network is done with much less loss (<0.1%) than in a power network (8%). The use of the existing natural gas pipelines for hydrogen was studied by NaturalHy.

## The Power-to-ammonia Concept

The power-to-ammonia concept offers a carbon-free energy storage route with a diversified application palette. At times when there is a surplus of renewable electricity, it can be converted to ammonia locally using small-scale plants. Existing technology can be used to produce ammonia by splitting water into hydrogen and oxygen with the help of electricity, then using high temperature and pressure to convert the hydrogen plus nitrogen from the air into ammonia. Ammonia is similar to propane when stored in liquid form, unlike Hydrogen which is difficult to liquefy and store cryogenicly at −252.882 °C.

Just like natural gas, the produced and stored ammonia can be used by energy companies at any time as fuel for electricity generation. Ammonia can be stored as a liquid; a standard tank of 60,000 m3 contains about 211 GWh of energy, equivalent to the annual production of roughly 30 wind turbines on land. Ammonia can be burned cleanly: water and nitrogen are released, but no CO2 and little or no nitrogen oxides. Ammonia can be further used as a flexible chemical as a fertilizer, chemical commodity, de-NOx agent and energy carrier. Given this flexibility of usage, and given the fact that the supporting infrastructure for the transport, distribution and usage of ammonia is already in place, it makes ammonia a good candidate to be a large-scale, non-carbon, energy carrier of the future.

## Hydroelectricity

## Pumped Water

Mingtan Pumped Storage Hydro Power Plant dam in Nantou, Taiwan

In 2008 world pumped storage generating capacity was 104 GW, while other sources claim 127 GW, which comprises the vast majority of all types of grid electric storage – all other types combined are some hundreds of MW.

In many places, pumped storage hydroelectricity is used to even out the daily generating load, by pumping water to a high storage reservoir during off-peak hours and weekends, using the excess base-load capacity from coal or nuclear sources. During peak hours, this water can be used for hydroelectric generation, often as a high value rapid-response reserve to cover transient peaks in demand. Pumped storage recovers about 70% to 85% of the energy consumed, and is currently the most cost effective form of mass power storage. The chief problem with pumped storage is that it usually requires two nearby reservoirs at considerably different heights, and often requires considerable capital expenditure.

Pumped water systems have high dispatchability, meaning they can come on-line very quickly, typically within 15 seconds, which makes these systems very efficient at soaking up variability in electrical *demand* from consumers. There is over 90 GW of pumped storage in operation around the world, which is about 3% of *instantaneous* global generation capacity. Pumped water storage systems, such as the Dinorwig storage system in Britain, hold five or six hours of generating capacity, and are used to smooth out demand variations.

Another example is the 1836 MW Tianhuangping Pumped-Storage Hydro Plant in China, which has a reservoir capacity of eight million cubic meters (2.1 billion U.S. gallons or the volume of water over Niagara Falls in 25 minutes) with a vertical distance of 600 m (1970 feet). The reservoir can provide about 13 GW·h of stored gravitational potential energy (convertible to electricity at about 80% efficiency), or about 2% of China's daily electricity consumption.

A new concept in pumped-storage is utilizing wind energy or solar power to pump water. Wind turbines or solar cells that direct drive water pumps for an energy storing wind or solar dam can make this a more efficient process but are limited. Such systems can only increase kinetic water volume during windy and daylight periods.

## Hydroelectric Dams

Fetsui hydroelectric dam in New Taipei, Taiwan

Hydroelectric dams with large reservoirs can also be operated to provide peak generation at times of peak demand. Water is stored in the reservoir during periods of low demand and released through the plant when demand is higher. The net effect is the same as pumped storage, but without the pumping loss. Depending on the reservoir capacity the plant can provide daily, weekly, or seasonal load following.

Many existing hydroelectric dams are fairly old (for example, the Hoover Dam was built in the 1930s), and their original design predated the newer intermittent power sources such as wind and solar by decades. A hydroelectric dam originally built to provide baseload power will have its generators sized according to the average flow of water into the reservoir. Uprating such a dam with additional generators increases its peak power output capacity, thereby increasing its capacity to operate as a virtual grid energy storage unit. The United States Bureau of Reclamation reports an investment cost of $69 per kilowatt capacity to uprate an existing dam, compared to more than $400 per kilowatt for oil-fired peaking generators. While an uprated hydroelectric dam does not directly store excess energy from other generating units, it behaves equivalently by accumulating its own fuel – incoming river water – during periods of high output from other generating units. Functioning as a virtual grid storage unit in this way, the uprated dam is one of the most efficient forms of energy storage, because it has no pumping losses to fill its reservoir, only increased losses to evaporation and leakage.

A dam which impounds a large reservoir can store and release a correspondingly large amount of energy, by controlling river outflow and raising or lowering its reservoir level a few meters. Limitations do apply to dam operation, their releases are commonly subject to government regulated water rights to limit downstream effect on rivers. For example, there are grid situations where baseload thermal plants, nuclear or wind turbines are already producing excess power at night, dams are still required to release enough water to maintain adequate river levels, whether electricity is generated or not. Conversely there's a limit to peak capacity, which if excessive could cause a river to flood for a few hours each day.

## Superconducting Magnetic Energy

Superconducting magnetic energy storage (SMES) systems store energy in the magnetic field created by the flow of direct current in a superconducting coil which has been cryogenically cooled to a temperature below its superconducting critical temperature. A typical SMES system includes three parts: superconducting coil, power conditioning system and cryogenically cooled refrigerator. Once the superconducting coil is charged, the current will not decay and the magnetic energy can be stored indefinitely. The stored energy can be released back to the network by discharging the coil. The power conditioning system uses an inverter/rectifier to transform alternating current (AC) power to direct current or convert DC back to AC power. The inverter/rectifier accounts for about 2–3% energy loss in each direction. SMES loses the least amount of electricity in the energy storage process compared to other methods of storing energy. SMES systems are highly efficient; the round-trip efficiency is greater than 95%. The high cost of superconductors is the primary limitation for commercial use of this energy storage method.

Due to the energy requirements of refrigeration, and the limits in the total energy able to be stored, SMES is currently used for short duration energy storage. Therefore, SMES is most commonly devoted to improving power quality. If SMES were to be used for utilities it would be a diurnal storage device, charged from base load power at night and meeting peak loads during the day.

Superconducting magnetic energy storage technical challenges are yet to be solved for it to become practical.

## Thermal

In Denmark the direct storage of electricity is perceived as too expensive for very large scale usage, albeit significant usage is made of existing Norwegian Hydro. Instead, the use of existing hot water storage tanks connected to district heating schemes, heated by either electrode boilers or heat pumps, is seen as a preferable approach. The stored heat is then transmitted to dwellings using district heating pipes.

Molten salt is used to store heat collected by a solar power tower so that it can be used to generate electricity in bad weather or at night.

Off-peak electricity can be used to make ice from water, and the ice can be stored. The stored ice can be used to cool the air in a large building which would have normally used electric AC, thereby shifting the electric load to off-peak hours. On other systems stored ice is used to cool the intake air of a gas turbine generator, thus increasing the on-peak generation capacity and the on-peak efficiency.

A Pumped Heat Electricity Storage system uses a highly reversible heat engine/heat pump to pump heat between two storage vessels, heating one and cooling the other. The UK-based engineering company Isentropic that is developing the system claims a potential electricity-in to electricity-out round-trip efficiency of 72–80%.

## Gravitational Potential Energy Storage with Solid Masses

According to Scientific American, ski lifts and trains tracks are among places being considered to store energy by moving heavy objects up or down inclines.

## Economics

The levelized cost of storing electricity depends highly on storage type and purpose; as subsecond-scale frequency regulation, minute/hour-scale peaker plants, or day/week-scale season storage.

Using battery storage is said to be U\$0.12-0.17 per kWh.

Generally speaking, energy storage is economical when the marginal cost of electricity varies more than the costs of storing and retrieving the energy plus the price of energy lost in the process. For instance, assume a pumped-storage reservoir can pump to its upper reservoir a volume of water capable of producing 1,200 MW·h after all losses are factored in (evaporation and seeping in the reservoir, efficiency losses, etc.). If the marginal cost of electricity during off-peak times is \$15 per MW·h, and the reservoir operates at 75% efficiency (i.e., 1,600 MW·h are consumed and 1,200 MW·h of energy are retrieved), then the total cost of filling the reservoir is \$24,000. If all of the stored energy is sold the following day during peak hours for an average \$40 per MW·h, then the reservoir will see revenues of \$48,000 for the day, for a gross profit of \$24,000.

However, the marginal cost of electricity varies because of the varying operational and fuel costs of different classes of generators. At one extreme, base load power plants such as coal-fired power plants and nuclear power plants are low marginal cost generators, as they have high capital and maintenance costs but low fuel costs. At the other extreme, peaking power plants such as gas turbine natural gas plants burn expensive fuel but are cheaper to build, operate and maintain. To minimize the total operational cost of generating power, base load generators are dispatched most of the time, while peak power generators are dispatched only when necessary, generally when energy demand peaks. This is called "economic dispatch".

Demand for electricity from the world's various grids varies over the course of the day and from season to season. For the most part, variation in electric demand is met by varying the amount of electrical energy supplied from primary sources. Increasingly, however, operators are storing lower-cost energy produced at night, then releasing it to the grid during the peak periods of the day when it is more valuable. In areas where hydroelectric dams exist, release can be delayed until demand is greater; this form of storage is common and can make use of existing reservoirs. This is not storing "surplus" energy produced elsewhere, but the net effect is the same – although without the efficiency losses. Renewable supplies with variable production, like wind and solar power, tend to increase the net variation in electric load, increasing the opportunity for grid energy storage.

It may be more economical to find an alternative market for unused electricity, rather than try and store it. High Voltage Direct Current allows for transmission of electricity, losing only 3% per 1000 km.

The United States Department of Energy's International Energy Storage Database provides a free list of grid energy storage projects, many of which show funding sources and amounts.

## Load Leveling

The demand for electricity from consumers and industry is constantly changing, broadly within the following categories:

- Seasonal (during dark winters more electric lighting and heating is required, while in other climates hot weather boosts the requirement for air conditioning)

- Weekly (most industry closes at the weekend, lowering demand)

- Daily (such as the morning peak as offices open and air conditioners get switched on)

- Hourly (one method for estimating television viewing figures in the United Kingdom is to measure the power spikes during advertisement breaks or after programmes when viewers go to switch a kettle on )

- Transient (fluctuations due to individual's actions, differences in power transmission efficiency and other small factors that need to be accounted for)

There are currently three main methods for dealing with changing demand:

- Electrical devices generally having a working voltage range that they require, commonly 110–120 V or 220–240 V. Minor variations in load are automatically smoothed by slight variations in the voltage available across the system.

- Power plants can be run below their normal output, with the facility to increase the amount they generate almost instantaneously. This is termed 'spinning reserve'.

- Additional generation can be brought online. Typically, these would be hydroelectric or gas turbines, which can be started in a matter of minutes.

The problem with standby gas turbines is higher costs, expensive generating equipment is unused much of the time. Spinning reserve also comes at a cost, plants run below maximum output are usually less efficient. Grid energy storage is used to shift generation from times of peak load to off-peak hours. Power plants are able to run at their peak efficiency during nights and weekends.

Supply-demand leveling strategies may be intended to reduce the cost of supplying peak power or to compensate for the intermittent generation of wind and solar power.

## Energy Demand Management

In order to keep the supply of electricity consistent and to deal with varying electrical loads it is necessary to decrease the difference between generation and demand. If this is done by changing loads it is referred to as demand side management (DSM). For decades, utilities have sold off-peak power to large consumers at lower rates, to encourage these users to shift their loads to off-peak hours, in the same way that telephone companies do with individual customers. Usually, these time-dependent prices are negotiated ahead of time. In an attempt to save more money, some utilities are experimenting with selling electricity at minute-by-minute spot prices, which allow those users with monitoring equipment to detect demand peaks as they happen, and shift demand to save both the user and the utility money. Demand side management can be manual or automatic and is not limited to large industrial customers. In residential and small business applications, for example, appliance control modules can reduce energy usage of water heaters, air conditioning units, refrigerators, and other devices during these periods by turning them off for some portion of the peak demand time or by reducing the power that they draw. Energy demand management includes more than reducing overall energy use or shifting loads to off-peak hours. A particularly effective method of energy demand management involves encouraging electric consumers to install more energy efficient equipment. For example, many utilities give rebates for the purchase of insulation, weatherstripping, and appliances and light bulbs that are energy efficient. Some utilities subsidize the purchase of geothermal heat pumps by their customers, to reduce electricity demand during the summer months by making air conditioning up to 70% more efficient, as well as to reduce the winter electricity demand compared to conventional air-sourced heat pumps or resistive heating. Companies with factories and large buildings can also install such products, but they can also buy energy efficient industrial equipment, like boilers, or use more efficient processes to produce products. Companies may get incentives like rebates or low interest loans from utilities or the government for the installation of energy efficient industrial equipment. Facilities may shift their demand by enlisting a third party to provide Energy Storage as a Service (ESaaS).

## Portability

This is the area of greatest success for current energy storage technologies. Single-use and rechargeable batteries are ubiquitous, and provide power for devices with demands as varied as digital watches and cars. Advances in battery technology have generally been slow, however, with much of the advance in battery life that consumers see being attributable to efficient power management rather than increased storage capacity. Portable consumer electronics have benefited greatly from size and power reductions associated with Moore's law. Unfortunately, Moore's law does not apply to hauling people and freight; the underlying energy requirements for transportation remain much higher than for information and entertainment applications. Battery capacity has become an issue as pressure grows for alternatives to internal combustion engines in cars, trucks, buses, trains, ships, and aeroplanes. These uses require far more energy density (the amount of energy stored in a given volume or weight) than current battery technology can deliver. Liquid hydrocarbon fuel (such as gasoline/petrol and diesel), as well as alcohols (methanol, ethanol, and butanol) and lipids (straight vegetable oil, biodiesel) have much higher energy densities.

There are synthetic pathways for using electricity to reduce carbon dioxide and water to liquid hydrocarbon or alcohol fuels. These pathways begin with electrolysis of water to generate hydrogen, and then reducing carbon dioxide with excess hydrogen in variations of the reverse water gas shift reaction. Non-fossil sources of carbon dioxide include fermentation plants and sewage treatment plants. Converting electrical energy to carbon-based liquid fuel has potential to provide portable energy storage usable by the large existing stock of motor vehicles and other engine-driven equipment, without the difficulties of dealing with hydrogen or another exotic energy carrier. These synthetic pathways may attract attention in connection with attempts to improve energy security in nations that rely on imported petroleum, but have or can develop large sources of renewable or nuclear electricity, as well as to deal with possible future declines in the amount of petroleum available to import.

Because the transport sector uses the energy from petroleum very inefficiently, replacing petroleum with electricity for mobile energy will not require very large investments over many years.

## Reliability

Virtually all devices that operate on electricity are adversely affected by the sudden removal of their power supply. Solutions such as UPS (uninterruptible power supplies) or backup generators are available, but these are expensive. Efficient methods of power storage would allow for devices to have a built-in backup for power cuts, and also reduce the impact of a failure in a generating station. Examples of this are currently available using fuel cells and flywheels.

# Compressed Air Energy Storage

A pressurized air tank used to start a diesel generator set in Paris Metro

Compressed air energy storage (CAES) is a way to store energy generated at one time for use at another time using compressed air. At utility scale, energy generated during periods of low energy demand (off-peak) can be released to meet higher demand (peak load) periods. Small scale systems have long been used in such applications as propulsion of mine locomotives. Large scale applications must conserve the heat energy associated with compressing air; dissipating heat lowers the energy efficiency of the storage system.

## Types

Compression of air creates heat; the air is warmer after compression. Expansion removes heat. If no extra heat is added, the air will be much colder after expansion. If the heat generated during compression can be stored and used during expansion, the efficiency of the storage improves considerably. There are three ways in which a CAES system can deal with the heat. Air storage can be adiabatic, diabatic, or isothermal.

## Adiabatic

Adiabatic storage continues to keep the heat produced by compression and returns it to the air as it is expanded to generate power. This is a subject of ongoing study, with no utility scale plants as of 2015, but a German project ADELE is planning to bring a demonstration plant (360 MWh storage capacity) into service in 2016. The theoretical efficiency of adiabatic storage approaches 100% with perfect insulation, but in practice round trip efficiency is expected to be 70%. Heat can be stored in a solid such as concrete or stone, or more likely in a fluid such as hot oil (up to 300 °C) or molten salt solutions (600 °C).

## Diabatic

Diabatic storage dissipates much of the heat of compression with intercoolers (thus approaching isothermal compression) into the atmosphere as waste; essentially wasting, thereby, the renewable energy used to perform the work of compression. Upon removal from storage, the temperature of this compressed air is *the one indicator* of the amount of stored energy that remains in this air. Consequently, if the air temperature is low for the energy recovery process, the air must be substantially re-heated prior to expansion in the turbine to power a generator. This reheating can be accomplished with a natural gas fired burner for utility grade storage or with a heated metal mass. As recovery is often most needed when renewable sources are quiescent, fuel must be burned to make up for the *wasted* heat. This degrades the efficiency of the storage-recovery cycle; and while this approach is relatively simple, the burning of fuel adds to the cost of the recovered electrical energy and compromises the ecological benefits associated with most renewable energy sources. Nevertheless, this is thus far the only system which has been implemented commercially.

The McIntosh, Alabama CAES plant requires 2.5 MJ of electricity and 1.2 MJ lower heating value (LHV) of gas for each MJ of energy output, corresponding to an energy recovery efficiency of about 27%. A General Electric 7FA 2x1 combined cycle plant, one of the most efficient natural gas plants in operation, uses 1.85 MJ (LHV) of gas per MJ generated, a 54% thermal efficiency.

## Isothermal

Isothermal compression and expansion approaches attempt to maintain operating temperature by constant heat exchange to the environment. They are only practical for low power levels, without very effective heat exchangers. The theoretical efficiency of isothermal energy storage approaches 100% for perfect heat transfer to the environment. In practice neither of these perfect thermodynamic cycles is obtainable, as some heat losses are unavoidable.

## Near Isothermal

Near isothermal compression (and expansion) is a process in which air is compressed in very close proximity to a large incompressible thermal mass such as a heat absorbing and releasing structure (HARS) or a water spray. A HARS is usually made up of a series of parallel fins. As the air is compressed the heat of compression is rapidly transferred to the thermal mass, so the gas temperature is stabilised. An external cooling circuit is then used to maintain the temperature of the thermal mass. The isothermal efficiency (Z) is a measure of where the process lies between an adiabatic and isothermal process. If the efficiency is 0%, then it is totally adiabatic; with an efficiency of 100%, it is totally isothermal. Typically with a near isothermal process an efficiency of 90-95% can be expected.

## Other

One implementation of isothermal CAES uses high, medium and low pressure pistons in series, with each stage followed by an airblast venturi pump that draws ambient air over an air-to-air (or air-to-seawater) heat exchanger between each expansion stage. Early compressed air torpedo designs used a similar approach, substituting seawater for air. The venturi warms the exhaust of the preceding stage and admits this preheated air to the following stage. This approach was widely adopted in various compressed air vehicles such as H. K. Porter, Inc.'s mining locomotives and trams. Here the heat of compression is effectively stored in the atmosphere (or sea) and returned later on.

## Compressors and Expanders

Compression can be done with electrically powered turbo-compressors and expansion with turbo 'expanders' or air engines driving electrical generators to produce electricity.

## Storage

The storage system of a CAES (Compressed Air Energy Storage) is one of the most interesting characteristics of this technology, and it is strictly related to its economic feasibility, energy density and flexibility. There are a few categories of air storage vessels, based on the thermodynamic conditions of the storage, and on the technology chosen:

1.  Constant Volume Storage (Solution mined caverns, aboveground vessels, aquifers, automotive applications, etc.)

2.  Constant Pressure Storage (Underwater pressure vessels, Hybrid Pumped Hydro - Compressed Air Storage)

## Constant Volume Storage

This storage system uses a chamber with rigid boundaries to store large amounts of air. This means from a thermodynamic point of view, that this system is a Constant Volume and Variable Pressure system. This causes some operational problems to the compressors and turbines operating on them, so the pressure variations have to be kept below a certain limit, as do the stresses induced on the storage vessels.

The storage vessel is often an underground cavern created by solution mining (salt is dissolved in water for extraction) or by utilizing an abandoned mine; use of porous rock formations(rocks which have holes through which liquid or air can pass) such as those in which reservoirs of natural gas are found has also been studied.

In some cases also an above ground pipeline was tested as a storage system, giving some good results. Obviously the cost of the system is higher, but it can be placed wherever the designer chooses, while an underground system needs some particular geologic formations (salt domes, aquifers, depleted gas mines..etc.).

## Constant Pressure Storage

In this case the storage vessel is kept at a constant pressure, while the gas is contained in a variable volume vessel. Many types of storage vessel have been proposed, but the operating conditions follow the same principle, the storage vessel is positioned hundreds of meters underwater, the hydrostatic pressure of the water column above the storage vessel allows to keep the pressure to the desired level.

This configuration allows to:

- Improve the energy density of the storage system, because all the air contained can be used (the pressure is constant in all charge conditions, full or empty, the pressure is the same, so the turbine has no problem exploiting it, while with constant volume systems after a while the pressure goes below a safety limit and the system needs to stop)

- Improve the efficiency of the turbomachinery, which will work under constant inlet conditions.

- Opens to the use of different geographic locations for the positioning of the CAES plant (coastal lines, floating platforms, etc.)

On the other hand, the cost of this storage system is higher, due to the need of positioning the storage vessel on the bottom of the chosen water reservoir (often the sea or the ocean) and due to the cost of the vessel itself.

Plants operate on a daily cycle, charging at night and discharging during the day. Heating of the compressed air using natural gas or geothermal heat to increase the amount of energy being extracted has been studied by the Pacific Northwest National Laboratory.

Compressed air energy storage can also be employed on a smaller scale such as exploited by air cars and air-driven locomotives, and can use high-strength carbon-fiber air storage tanks. In order to retain the energy stored in compressed air, this tank should be thermally isolated from the environment; else, the energy stored will escape under the form of heat since compressing air raises its temperature.

## History

## Transmission

Citywide compressed air energy systems have been built since 1870. Cities such as Paris, France; Birmingham, England; Dresden, Rixdorf and Offenbach, Germany and Buenos Aires, Argentina installed such systems. Victor Popp constructed the first systems to power clocks by sending a pulse of air every minute to change their pointer arms. They quickly evolved to deliver power to homes and industry. As of 1896, the Paris system had 2.2 MW of generation distributed at 550 kPa in 50 km of air pipes for motors in light and heavy industry. Usage was measured by meters. The systems were the main source of house-delivered energy in those days and also powered the machines of dentists, seamstresses, printing facilities and bakeries.

## Storage

- 1978 – The first utility-scale compressed air energy storage project was the 290 megawatt Huntorf plant in Germany using a salt dome.

- 1991 – A 110 megawatt plant with a capacity of 26 hours was built in McIntosh, Alabama (1991). The Alabama facility's $65 million cost works out to $590 per kW of generation capacity and about $23 per kW-hr of storage capacity, using a 19 million cubic foot solution mined salt cavern to store air at up to 1100 psi. Although the compression phase is approximately 82% efficient, the expansion phase requires combustion of natural gas at one third the rate of a gas turbine producing the same amount of electricity.

- December, 2012 – General Compression completes construction of a 2 MW near-isothermal CAES project in Gaines, TX; the world's third CAES project. The project uses no fuel.

## Projects

- Huntorf plant in Germany (290 MW) non-adiabatic.

- McIntosh plant in Alabama (226 MW) non-adiabatic.

- November 2009 – The US Department of Energy awards $24.9 million in matching funds for phase one of a 300 MW, $356 million Pacific Gas and Electric CAES installation utilizing a saline porous rock formation being developed near Bakersfield in Kern County, California. Goals of the project are to build and validate an advanced design.

- December, 2010 – The US Department of Energy provides $29.4 million in funding to conduct preliminary work on a 150 MW salt-based CAES project being developed by Iberdrola USA in Watkins Glen, New York. The goal is to incorporate smart grid technology to balance renewable intermittent energy sources.

- 2013 – The first adiabatic CAES project, a 200 megawatt facility called ADELE, was planned for construction in Germany. This project has been delayed for undisclosed reasons until at least 2016.

- 2016 (projected) – Apex has planned a CAES plant for Anderson County, Texas to go online in 2016. This project has been delayed and will not go into operation until July 2017

- 2017 (projected) - Storelectric Ltd is planning to build a 40 MW 100% renewable energy pilot plant in Cheshire, UK, with 800 MWh storage capacity. "This would be 20 times larger than any 100% renewable energy CAES built so far, representing a step-change in the storage industry." according to their website.

- Larne, Northern Ireland - a 330 MW CAES project to solution-mine two caverns in a salt deposit, supported by EU with €90 million.

- Europen Union-funded RICAS 2020 (adiabatic) project in Austria uses crushed rock to store heat from the compression process to improve efficiency. The system was expected to achieve 70-80% efficiency.

## Storage Thermodynamics

In order to achieve a near thermodynamic reversible process so that most of the energy is saved in the system and can be retrieved, and losses are kept negligible, a near reversible isothermal process or an isentropic process is desired.

## Isothermal Storage

In an isothermal compression process, the gas in the system is kept at a constant temperature throughout. This necessarily requires exchange of heat with the gas, otherwise the temperature would rise during charging and drop during discharge. This heat exchange can be achieved by heat exchangers (intercooling) between subsequent stages in the compressor, regulator and tank. To avoid wasted energy, the intercoolers must be optimised for high heat transfer and low pressure drop. Smaller compressors can approximate isothermal compression even without intercooling, due to the relatively high ratio of surface area to volume of the compression chamber and the resulting improvement in heat dissipation from the compressor body itself.

When one obtains perfect isothermal storage (and discharge), the process is said to be "reversible". This requires that the heat transfer between the surroundings and the gas occur over an infinitesimally small temperature difference. In that case, there is no exergy loss in the heat transfer process, and so the compression work can be completely recovered as expansion work: 100% storage efficiency. However, in practice, there is always a temperature difference in any heat transfer process, and so all practical energy storage obtains efficiencies lower than 100%.

To estimate the compression/expansion work in an isothermal process, it may be assumed that the compressed air obeys the ideal gas law,

$$pV = nRT = \text{constant}.$$

From a process from an initial state $A$ to a final state $B$, with absolute temperature $T = T_A = T_B$ constant, one finds the work required for compression (negative) or done by the expansion (positive), to be

$$W_{A \to B} = \int_{V_A}^{V_B} p \, dV = \int_{V_A}^{V_B} \frac{nRT}{V} \, dV = nRT \int_{V_A}^{V_B} \frac{1}{V} \, dV$$

$$= nRT(\ln V_B - \ln V_A) = nRT \ln \frac{V_B}{V_A} = nRT \ln \frac{p_A}{p_B} = p_A V_A \ln \frac{p_A}{p_B}$$

where $pV = p_A V_A = p_B V_B$, and so, $\dfrac{V_B}{V_A} = \dfrac{p_A}{p_B}$. Here, $p$ is the absolute pressure, $V$ is the volume of the vessel, $n$ is the amount of substance of gas (mol) and $R$ is the ideal gas constant.

If there is a constant pressure outside of the vessel which is equal to the starting pressure $p_A$, the positive work of the outer pressure reduces the exploitable energy (negative value). This adds a term to the equation above:

$$W_{A \to B} = p_A V_A \ln \frac{p_A}{p_B} + (V_A - V_B)p_A = p_A V_A \ln \frac{p_A}{p_B} + (p_B - p_A)V_B$$

*Example*

How much energy can be stored in a 1 m³ storage vessel at a pressure of 70 bars (7.0 MPa), if the ambient pressure is 1 bar (0.10 MPa). In this case, the process work is

$$W = p_B v_B \ln \frac{p_A}{p_B} + (p_B - p_A)V_B =$$

= 7.0 MPa × 1 m³ × ln(0.1 MPa/7.0 MPa) + (7.0 MPa - 0.1 MPa) x 1 m³ = -22.8 MJ (equivalently 6.33 KWh).

The negative sign means that work is done on the gas by the surroundings. Process irreversibilities (such as in heat transfer) will result in less energy being recovered from the expansion process than is required for the compression process. If the environment is at a constant temperature, for example, the thermal resistance in the intercoolers will mean that the compression occurs at a temperature somewhat higher than the ambient temperature, and the expansion will occur at a temperature somewhat lower than ambient temperature. So a perfect isothermal storage system is impossible to achieve.

## Adiabatic (Isentropic) Storage

An adiabatic process is one where there is no heat transfer between the fluid and the surroundings: the system is insulated against heat transfer. If the process is furthermore internally reversible (smooth, slow and frictionless, to the ideal limit) then it will additionally be isentropic.

An adiabatic storage system does away with the intercooling during the compression process, and simply allows the gas to heat up during compression, and likewise to cool down during expansion. This is attractive, since the energy losses associated with the heat transfer are avoided, but the downside is that the storage vessel must be insulated against heat loss. It should also be mentioned that real compressors and turbines are not isentropic, but instead have an isentropic efficiency of around 85%, with the result that round-trip storage efficiency for adiabatic systems is also considerably less than perfect.

## Large Storage System Thermodynamics

Energy storage systems often use large underground caverns. This is the preferred system design, due to the very large volume, and thus the large quantity of energy that can be stored with only a small pressure change. The cavern space can be easily insulated, compressed adiabatically with little temperature change (approaching a reversible isothermal system) and heat loss (approaching an isentropic system). This advantage is in addition to the low cost of constructing the gas storage system, using the underground walls to assist in containing the pressure.

Recently there have been developed undersea insulated air bags, with similar thermodynamic properties to large underground cavern storage.

## Practical Constraints in Transportation

In order to use air storage in vehicles or aircraft for practical land or air transportation, the energy storage system must be compact and lightweight. Energy density and specific energy are the engineering terms that define these desired qualities.

## Specific Energy, Energy Density and Efficiency

Compressing air heats it and expanding it cools it. Therefore, practical air engines require heat exchangers in order to avoid excessively high or low temperatures and even so don't reach ideal constant temperature conditions, or ideal thermal insulation.

Nevertheless, as stated above, it is useful to describe the maximum energy storable using the isothermal case, which works out to about 100 kJ/m³ [ $\ln(P_A/P_B)$ ].

Thus if 1.0 m³ of air from the atmosphere is very slowly compressed into a 5 L bottle at 20 MPa (200 bar), the potential energy stored is 530 kJ. A highly efficient air motor can transfer this into kinetic energy if it runs very slowly and manages to expand the air from its initial 20 MPa pressure down to 100 kPa (bottle completely "empty" at atmospheric pressure). Achieving high efficiency is a technical challenge both due to heat loss to the ambient and to unrecoverable internal gas heat. If the bottle above is emptied to 1 MPa, the extractable energy is about 300 kJ at the motor shaft.

A standard 20 MPa, 5 L steel bottle has a mass of 7.5 kg, a superior one 5 kg. High-tensile strength fibers such as carbon-fiber or Kevlar can weigh below 2 kg in this size, consistent with the legal safety codes. One cubic meter of air at 20 °C has a mass of 1.204 kg at standard temperature and pressure. Thus, *theoretical* specific energies are from roughly 70 kJ/kg at the motor shaft for a plain steel bottle to 180 kJ/kg for an advanced fiber-wound one, whereas practical *achievable* specific energies for the same containers would be from 40 to 100 kJ/kg.

## Comparison with Batteries

Advanced fiber-reinforced bottles are comparable to the rechargeable lead-acid battery in terms of energy density. Batteries provide nearly constant voltage over their entire charge level, whereas the pressure varies greatly while using a pressure vessel from full to empty. It is technically challenging to design air engines to maintain high efficiency and sufficient power over a wide range of pressures. Compressed air can transfer power at very high flux rates, which meets the principal acceleration and deceleration objectives of transportation systems, particularly for hybrid vehicles.

Compressed air systems have advantages over conventional batteries including longer lifetimes of pressure vessels and lower material toxicity. Newer battery designs such as those based on Lithium Iron Phosphate chemistry suffer from neither of these problems. Compressed air costs are potentially lower; however advanced pressure vessels are costly to develop and safety-test and at present are more expensive than mass-produced batteries.

As with electric storage technology, compressed air is only as "clean" as the source of the energy that it stores. Life cycle assessment addresses the question of overall emissions from a given energy storage technology combined with a given mix of generation on a power grid.

## Safety

As with most technologies, compressed air has safety concerns, mainly catastrophic tank rupture. Safety codes make this a rare occurrence at the cost of higher weight and additional safety features such as pressure relief valves. Codes may limit the legal working pressure to less than 40% of the rupture pressure for steel bottles (safety factor of 2.5), and less than 20% for fiber-wound bottles (safety factor of 5). Commercial designs adopt the ISO 11439 standard. High pressure bottles are fairly strong so that they generally do not rupture in vehicle crashes.

## History

Air engines have been used since the 19th century to power mine locomotives, pumps, drills and trams, via centralized, city-level, distribution. Racecars use compressed air to start their internal combustion engine (ICE), and large Diesel engines may have starting pneumatic motors.

## Engine

A compressed air engine uses the expansion of compressed air to drive the pistons of an engine, turn the axle, or to drive a turbine.

The following methods can increase efficiency:

- A continuous expansion turbine at high efficiency

- Multiple expansion stages

- Use of waste heat, notably in a hybrid heat engine design

- Use of environmental heat

A highly efficient arrangement uses high, medium and low pressure pistons in series, with each stage followed by an airblast venturi that draws ambient air over an air-to-air heat exchanger. This warms the exhaust of the preceding stage and admits this preheated air to the following stage. The only exhaust gas from each stage is cold air which can be as cold as −15 °C (5 °F); the cold air may be used for air conditioning in a car.

Additional heat can be supplied by burning fuel as in 1904 for Whitehead's torpedoes. This improves the range and speed available for a given tank volume at the cost of the additional fuel.

## Cars

Since about 1990 several companies have claimed to be developing compressed air cars, but none is available. Typically the main claimed advantages are: no roadside pollution, low cost, use of cooking oil for lubrication, and integrated air conditioning.

The time required to refill a depleted tank is important for vehicle applications. "Volume transfer" moves pre-compressed air from a stationary tank to the vehicle tank almost instantaneously. Alternatively, a stationary or on-board compressor can compress air on demand, possibly requiring several hours.

## Ships

Large marine diesel engines are started using compressed air, typically between 20 and 30 bar and stored in two or more large bottles, acting directly on the pistons via special starting valves to turn the crankshaft prior to beginning fuel injection. This arrangement is more compact and cheaper than an electric starter motor would be at such scales, and able to supply the necessary burst of extremely high power without placing a prohibitive load on the ship's electrical generators and distribution system. Compressed air is commonly also used, at lower pressures, to control the engine and act as the spring force acting on the cylinder exhaust valves, and to operate other auxiliary systems and power tools on board, sometimes including pneumatic PID controllers. One advantage of this approach is that in the event of an electrical blackout, ship systems powered by stored compressed air can continue functioning uninterrupted, and generators can be restarted without an electrical supply. Another is that pneumatic tools can be used in commonly wet environments without risk of electric shock.

## Hybrid Vehicles

While the air storage system offers a relatively low power density and vehicle range, its high efficiency is attractive for hybrid vehicles that use a conventional internal combustion engine as a main power source. The air storage can be used for regenerative braking and to optimize the cycle of the piston engine which is not equally efficient at all power/RPM levels.

Bosch and PSA Peugeot Citroën have developed a hybrid system that use hydraulics as a way to transfer energy to and from a compressed nitrogen tank. An up to 45% reduction in fuel consumption is claimed, corresponding to 2.9l/100 km (81 mpg, 69 g $CO_2$/km) on the NEDC cycle for a compact frame like Peugeot 208. The system is claimed to be much more affordable than competing electric and flywheel KERS systems and is expected on road cars by 2016.

## Types of Systems

### Hybrid Systems

Brayton cycle engines compress and heat air with a fuel suitable for an internal combustion engine. For example, natural gas or biogas heat compressed air, and then a conventional gas turbine engine or the rear portion of a jet engine expands it to produce work.

Compressed air engines can recharge an electric battery. The apparently defunct Energine promoted its Pne-PHEV or Pneumatic Plug-in Hybrid Electric Vehicle-system.

### Existing Hybrid Systems

Huntorf, Germany in 1978, and McIntosh, Alabama, U.S. in 1991 commissioned hybrid power plants. Both systems use off-peak energy for air compression and burn natural gas in the compressed air during the power generating phase.

## Future Hybrid Systems

The Iowa Stored Energy Park (ISEP) will use aquifer storage rather than cavern storage. The displacement of water in the aquifer results in regulation of the air pressure by the constant hydrostatic pressure of the water. A spokesperson for ISEP claims, "you can optimize your equipment for better efficiency if you have a constant pressure." Power output of the McIntosh and Iowa systems is in the range of 2–300 MW.

Additional facilities are under development in Norton, Ohio. FirstEnergy, an Akron, Ohio electric utility obtained development rights to the 2,700 MW Norton project in November, 2009.

The RICAS2020 project attempts to use an abandoned mine for adiabatic CAES with heat recovery. The compression heat is stored in a tunnel section filled with loose stones, so the compressed air is nearly cool when entering the main pressure storage chamber. The cool compressed air regains the heat stored in the stones when released back through a surface turbine, leading to a higher overall efficiency.

## Lake or Ocean Storage

Deep water in lakes and the ocean can provide pressure without requiring high-pressure vessels or drilling into salt caverns or aquifers. The air goes into inexpensive, flexible containers such as plastic bags below in deep lakes or off sea coasts with steep drop-offs. Obstacles include the limited number of suitable locations and the need for high-pressure pipelines between the surface and the containers. Since the containers would be very inexpensive, the need for great pressure (and great depth) may not be as important. A key benefit of systems built on this concept is that charge and discharge pressures are a constant function of depth. Carnot inefficiencies can thereby be reduced in the power plant. Carnot efficiency can be increased by using multiple charge and discharge stages and using inexpensive heat sources and sinks such as cold water from rivers or hot water from solar ponds. Ideally, the system must be very clever—for example, by cooling air before pumping on summer days. It must be engineered to avoid inefficiency, such as wasteful pressure changes caused by inadequate piping diameter.

A nearly isobaric solution is possible if the compressed gas is used to drive a hydroelectric system. However, this solution requires large pressure tanks located on land (as well as the underwater air bags). Also, hydrogen gas is the preferred fluid, since other gases suffer from substantial hydrostatic pressures at even relatively modest depths (such as 500 meters).

E.ON, one of Europe's leading power and gas companies, has provided €1.4 million (£1.1 million) in funding to develop undersea air storage bags. Hydrostor in Canada is developing a commercial system of underwater storage "accumulators" for compressed air energy storage, starting at the 1 to 4 MW scale.

There is a plan for some type of compressed air energy storage in undersea caves by Northern Ireland.

## Near Isothermal

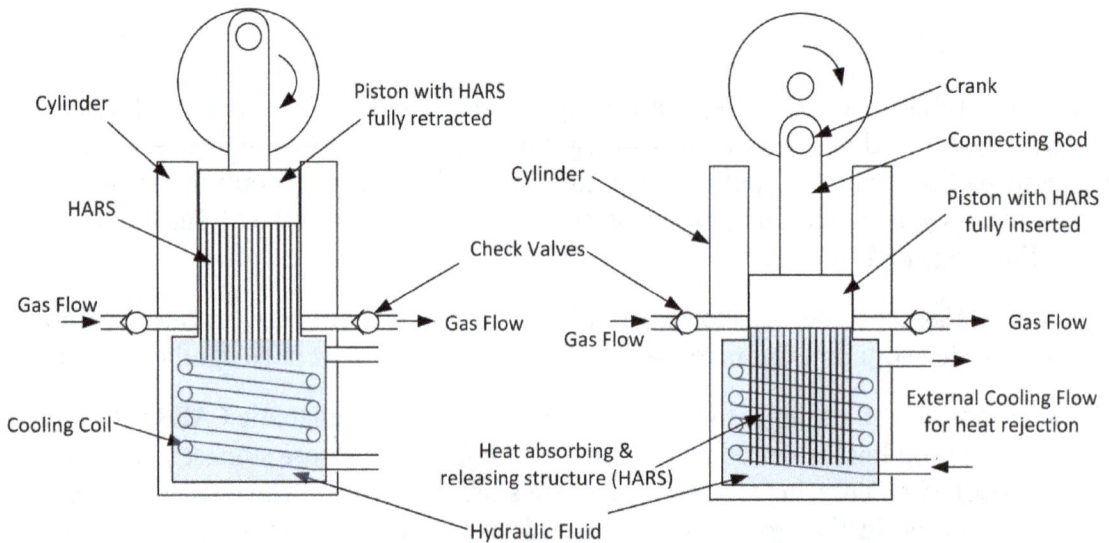

Schematic views of a near isothermal compressor and expander.
Left view with piston fully retracted right view with piston fully inserted.

A number of methods of near isothermal compression are being developed. Fluid Mechanics has a system with a heat absorbing and releasing structure (HARS) attached to a reciprocating piston. Light Sail inject a water spray into a reciprocating cylinder. SustainX use an air water foam mix inside a compressor. All these systems ensure that the air is compressed with high thermal diffusivity compared to the speed of compression. Typically these compressors can run at speeds up to 1000 rpm. To ensure high thermal diffusivity the average distance a gas molecule is from a heat absorbing surface is about 0.5mm. These near isothermal compressors can also be used as near isothermal expanders and are being developed to improve the round trip efficiency of CASE.

## Cryogenic Energy Storage

Cryogenic energy storage (CES) is the use of low temperature (cryogenic) liquids such as liquid air or liquid nitrogen as energy storage. Both cryogens have been used to power cars. The inventor Peter Dearman initially developed a liquid air car, and then used the technology he developed for grid energy storage. The technology is being piloted at a UK power station.

### History

A liquid air powered car called Liquid Air was built between 1899 and 1902 but it couldn't at the time compete in terms of efficiency with other engines. More recently, a liquid nitrogen vehicle was built. Peter Dearman, a garage inventor in Hertfordshire, UK who had initially developed a liquid air powered car, then put the technology to use as grid energy storage. The Dearman engine differs from former nitrogen engine designs in that the nitrogen is heated by combining it with the heat exchange fluid inside the cylinder of the engine.

## Grid Energy Storage

### Process

When it is cheaper (usually at night), electricity is used to cool air from the atmosphere to -195 °C using the Claude Cycle to the point where it liquefies. The liquid air, which takes up one-thousandth of the volume of the gas, can be kept for a long time in a large vacuum flask at atmospheric pressure. At times of high demand for electricity, the liquid air is pumped at high pressure into a heat exchanger, which acts as a boiler. Air from the atmosphere at ambient temperature, or hot water from an industrial heat source, is used to heat the liquid and turn it back into a gas. The massive increase in volume and pressure from this is used to drive a turbine to generate electricity.

### Efficiency

In isolation the process is only 25% efficient, but this is greatly increased (to around 50%) when used with a low-grade cold store, such as a large gravel bed, to capture the cold generated by evaporating the cryogen. The cold is re-used during the next refrigeration cycle.

Efficiency is further increased when used in conjunction with a power plant or other source of low-grade heat that would otherwise be lost to the atmosphere. Highview Power Storage claims an AC to AC round-trip efficiency of 70%, by using an otherwise waste heat source at 115 °C. The IMechE (Institution of Mechanical Engineers) agrees that these estimates for a commercial-scale plant are realistic. However this number was not checked or confirmed by independent professional institutions.

Currently surplus gaseous nitrogen is produced as a byproduct in the production of oxygen. Oxygen can be used in oxy-combustion coal power plants, enabling $CO_2$ capture and sequestration. This gaseous nitrogen can be liquefied by available liquefaction capacities for further use. Cryogenic distillation of air is currently the only commercially viable technology for large scale oxygen production.

### Pilot Plant

A 300 kW, 2.5MWh storage capacity pilot cryogenic energy system developed by researchers at the University of Leeds and Highview Power Storage, that uses liquid air (with the $CO_2$ and water removed as they would turn solid at the storage temperature) as the energy store, and low-grade waste heat to boost the thermal re-expansion of the air, operated at a 80MW biomass power station in Slough, UK, from 2010 until 2014 when it was relocated to the university of Birmingham. the efficiency is less than 15% because of low efficiency hardware components used, but the engineers are targeting an efficiency of about 60 percent for the next generation of CES based on operation experiences of this system.

The system is based on proven technology, used safely in many industrial processes, and does not require any particularly rare elements or expensive components to manufacture. Dr Tim Fox, the head of Energy at the IMechE says "it uses standard industrial components...., it will last for decades, and it can be fixed with a spanner."

In April 2014 the UK government announced it had given them £8 million to fund the next stage of the demonstration.

## Battery Storage Power Station

A battery storage power plant is a form of storage power plant, which uses batteries on an electrochemical basis for energy storage. Unlike common storage power plants, such as the pumped storage power plants with capacities up to 1000 MW, the benefits of battery storage power plants move in the range of a few kW up to the MW range - the largest installed systems (1/2017) reach capacities of up to 300 MWh. Battery storage power plants, like all storage power plants, primarily serve to cover peak load and in networks with insufficient control power and the grid stabilization. Small battery storage called solar batteries with few kWh storage capacity, are mostly in the private sector operated in conjunction with similarly sized photovoltaic systems to daytime bring revenue surpluses in yield poorer or unproductive hours in the evening or at night, and to strengthen their own consumption. Sometimes battery storage power stations are built with flywheel storage power systems in order to conserve battery power. Flywheels can handle rapid fluctuations better.

### Construction

Structurally battery storage power plants and uninterruptible power supplies (UPS) are comparable, although the former are larger. The batteries are housed for security in their own warehouses or in containers. As with a UPS, the problem is that electrochemical energy is stored or emitted in the form of direct current DC, while electric power networks are usually operated with Alternating current AC voltage. For this reason, additional inverters are needed to connect the battery storage power plants to the high voltage network. This kind of power electronics include GTO thyristors, commonly used in the high-voltage direct current transmission (HVDC). Various accumulator systems may be used depending on the power-to-energy ratio, the expected life time and, of course, the costs. In the 1980s, lead-acid batteries were used for the first battery-storage power plants. During the next few decades, nickel-cadmium and sodium-sulfur battery were increasingly used. Since 2010, more and more utility-scale battery storage plants rely on lithium-ion batteries thanks to the fast decrease in the cost of this technology, driven by the electric automotive industry. This is the case of the battery Park Schwerin, the battery storage in Dresden or the storage of BYD in Hong Kong. Since 2015, lithium-ion batteries are mainly used, some redox flow system have merged and lead-acid batteries are still used in small budget applications. There are numerous suppliers of large battery storage.

### Operating Characteristics

Since they do not require any mechanical movement, battery storage power plants allow extremely short control times and start times, in the range of few 10s of ms at full load. Thanks to that reactivity, they can shave power peaks in the range of minutes, but they can also dampen the fast oscillations (second) that appear when electric power networks are operated close to their maximum capacity. These instabilities consist in voltage fluctuations with periods of up to several 10 seconds and can soar in worst cases to high amplitudes, which can lead to regional blackouts. A battery storage power plants properly dimensioned can efficiently counteract these oscillations. Therefore, applications are found primarily in those regions where electrical power systems are operated at full capacity, causing a risk in the grid stability. Large storage plants (Na-S)can also be used in combination with (Na-S)intermittent renewable energy source in standalone hybrid micro-grids.

Some systems, operating at high temperature (Na-S) or using corrosive components are subject to failure even if they are not used (calendar ageing). Other technologies suffer from deterioration caused by charge-discharge cycles (cycle ageing), especially at high charging rates. These two types of ageing cause a loss of performance (capacity or voltage decrease), overheating and may eventually lead to critical failure (electrolyte leaks, fire, explosion). In order to prevent the loss of performance due to ageing, some batteries can undergo maintenance operation. For example, non-sealed Lead-acid batteries produce hydrogen and oxygen from the aqueous electrolyte when overcharged. The water has to be refilled regularly to avoid damage to the battery and the inflammable gases have to be vented out to avoid explosion risks. However, this maintenance has a cost and recent batteries, such as Li-Ion, are designed to have a long lifespan without maintenance. Therefore, most of the current systems are composed of securely sealed battery packs which are electronically monitored and replaced once their performance falls below a given threshold.

## Installation Examples

Below are some of the largest battery storage power plants are exemplified.

### Chino Battery Storage Project

The operated from 1988 to 1997 by the Southern California Edison in the Californian city Chino battery storage power station served primarily for grid stabilization and could be used by frequent power outages in the region as a static var compensator and black start of non-black bootable power plants. The plant had a peak power of 14 MW, which was, however, far too little for effective stabilization in the net of Southern California Edison, and a storage capacity of 40 MWh. The system consisted of 8,256 lead-acid batteries in eight strands, which were divided into two halls.

### Golden Valley Electric – Fairbanks

One of the largest and located with Stand 2010 operating system is operated by the Golden Valley Electric in Fairbanks. The power grid in Alaska is operated due to the large distances as stand-alone grid with no direct connection to neighboring North American interconnections within the North American Electric Reliability Corporation. The battery storage power plant with a maximum capacity of 27 MW is used to stabilize the grid, covering high peak and reactive power compensation. The plant was put into operation in 2003 and consists of 13,760 nickel-cadmium batteries in four strands. The NiCd cells are manufactured by Saft Groupe S.A., the inverters by ABB Group.

## BYD in Hongkong

The Chinese company BYD operates a battery banks with 40 MWh capacity and 20 MW maximum power in Hong Kong. The large storage is used to cushion load peaks in energy demand. Likewise, the storage can contribute to the frequency stabilization in the net. The battery is made up of a total of almost 60,000 individual lithium iron phosphate cells, each with 230 amp hour capacity. The project was started in October 2013, and went online in June 2014. The actual installation of the storage lasted three months. The use of price differences between loading and unloading by day and night electricity, an avoided grid expansion for peak loads and revenue for grid services such as Frequency stabilization enable economic operation without subsidies. Currently 3 locations for a 1,000 MW peak power to 200 MWh capacity storage power plant to be examined.

## Battery Storage Power Station Schwerin

In Schwerin, Germany the electricity supplier WEMAG operates a lithium-ion battery storage to compensate for short-term power fluctuations. Supplier of battery storage power station is the Berlin company Younicos. The South Korean company Samsung SDI supplied the lithium-ion cells. The memory has a capacity of 5 MWh and an output of 5 MW was in September 2014 in operation. The lithium-ion battery storage consists of 25,600 lithium manganese cells and is about five medium-voltage transformers with both the regional distribution connected as well with the nearby 380 kV high-voltage grid.

## Photovoltaic and Hybrid Power Plant

The existing photovoltaic power plant Alt Daber near Wittstock in Brandenburg, Germany received a battery storage of 2 MWh. A special feature is that this is a turnkey solution supplied and installed in containers, for immediate use on site without major construction work. The storage uses lead-acid batteries.

## Hybrid Battery Power Plant Braderup

Dince July 2014, the energy storage company Nord GmbH & Co. KG has been operating the largest hybrid batteries in Europe in Braderup (Schleswig-Holstein, Germany). The System consists of a lithium-ion battery storage (2 MW power 2 MWh storage) and a vanadium flow battery storage (330 kW power, 1 MWh storage capacity). The lithium-ion modules used are from Sony, the flow battery is made by Vanadis Power GmbH.

The storage system is connected to the local community wind park (18 MW installed capacity). Depending on wind strength and charging status of each battery a system developed by Bosch distributes the energy generated by the wind turbines to the right battery. Bosch is also responsible for project implementation and system integration. The hybrid battery is connected to the power grid by a ten-kilometer long underground cable. In case of a network congestion the batteries buffer the energy of the wind farm and feed it back into the grid at a convenient time later. With this method, a shutdown of wind turbines can be avoided during times of network congestion so that the energy of the wind is not wasted.

## Battery Storage Dresden

Stadtwerke Dresden, Germany (Drewag) have taken a battery storage with a peak power of 2 MW online on March 17, 2015. The costs amounted to 2.7 million euros. lithium polymer batteries are being used. The batteries including the control system are deployed in two 13 m long containers and can store a total of 2.7 MWh. The system is designed to compensate for peak power generation of a nearby solar plant.

## Battery Storage Feldheim

In Feldheim in Brandenburg, Germany, a battery storage with a capacity of 10 MW and a storage capacity of 6.5 MWh was put into operation in September, 2015. The project cost 12.8 million euros. The storage provides energy for the power grid to compensate for fluctuations caused by wind and solar power plants. The store is operated by the company Energiequelle.

## Evonik Battery Storage

Evonik is planning to build six battery storage power plants with a capacity of 15 MW to be put into operation in 2016 and 2017. They are to be situated in North Rhine-Westphalia, Germany at the power plant sites Herne, Lünen and Duisburg-Walsum and in Bexbach, Fenne and Weiher in the Saarland.

## Grand Ridge Power Plant in Illinois and Beech Ridge, West Virginia, USA

The largest grid storage batteries in the United States include the 31.5 MW battery at Grand Ridge Power plant in Illinois and the 31.5 MW battery at Beech Ridge, West Virginia. Both using lithium ion batteries.

## 400 MWh Southern California Edison Project

Under construction in 2015 is the 400 MWh (100 MW for 4 hours) Southern California Edison project. Developed by AES Energy it is a lithium-ion battery system. Southern California Edison found the prices for battery storage comparable with other electricity generators.

## 52 MWh Project on Kauai, Hawaii

Under construction (2015) is a 52 MWh project on Kauai, Hawaii to entirely time shift a 13 MW solar farm's output to the evening. The aim is to reduce dependence on fossil fuels on the island.

## 250 MWh Indonesia

At present (2/2016) is under construction a 250 MWh battery storage in Indonesia. There are about 500 villages in Indonesia which should be supplied, so far they depend on the power supply of petroleum. In past the prices fluctuated greatly and there was often power outages. Now the power will be generated through wind and solar power.

## Battery Storage Notrees, Texas, 36 MW

One battery is in Notrees, Texas (36 MW for 40 minutes using lead-acid batteries).

## Battery Storage with 13 MWh in Germany with Worn Batteries

A 13 MWh battery made of worn batteries from electric cars is being constructed in Germany, with an expected second life of 10 years, after which they will be recycled.

## 53 MWh in Ontario

In Ontario, Canada, a battery storage with 53 MWh capacity and 13 MW of power is established by the end of 2016. The Swiss battery manufacturer Leclanché supplies the batteries now. Deltro Energy Inc. will plan and build the plant. The order was placed by the network operator IESO. The energy storage are used to provide fast grid services, mainly for voltage and reactive power control. In Ontario and the surrounding area there are many wind and solar power plants, whereby the power supply varies widely.

## Storage in Southern England with Special Control

In southern England a battery storage with a capacity of 0.6 MWh and 0.3 MW power was installed for demonstration purposes, made up of 1400 lithium cells installed in a container, with a special feature being the control of the storage. Whereas usually a battery storage uses only revenue model, the provision of control energy, which is a very small market, this storage uses three revenue models. The storage has been installed next to a solar system. This way the solar system can be designed larger than the grid power actually permits in the first revenue model. The storage accepts a peak input of the solar system, thus avoiding the cost of a further grid expansion. The second model allows taking up peak input from the power grid and feeding it back to stabilize the grid when necessary. The third model is storing energy and feeding it into the grid at peak prices. The store received an award for top innovation.

## South Korea

Since January 2016 in South Korea three battery storage power plants are in operation. There are two new systems, a 24 MW system with 9 MWh and a 16 MW system with 6 MWh. These two uses batteries based on lithium-nickel-manganese-cobalt oxide and supplement a few months older system with 16MW and 5MWh whose batteries are based on lithium titanate oxide. Together the systems have a capacity of 56 MW and serve the South Korean utility company Korea Electric Power Corporation (KEPCO) for frequency regulation. The storage comes from the company Kokam. After completion in 2017, the system should have a power of 500 MW. The three already installed storage reduce annual fuel costs by an estimated 13 million US dollars, as well as cutting greenhouse gas emissions. Thus the saved fuel costs will exceed the cost of battery storage significantly.

## Storage for Aboriginal Community in Australia

An existing system in an Aboriginal community in Australia consisting of a combination photovoltaic system and diesel generator will be extended by a lithium-ion battery to a hybrid system. The battery has a capacity of about 2 MWh and a power of 0.8 MW. The batteries store the excess solar power and take over the previously network-forming functions such as network management and network stabilization of diesel generators. Thus, the diesel generators can be switched off during the day, which leads to cost reduction. Moreover, the share of renewable energy rises in the hybrid system significantly. The system is part of a plan to transform the energy systems of indigenous communities in Australia.

## Storage for Azores Island of Graciosa

On the Azores island of Graciosa a 3.2 MWh storage was installed. Along with a 1 MW photovoltaic plant and a 4.5 MW wind farm, the island is almost completely independent of the previously used diesel generators. The old power plant only serves as a backup system in the event that power from solar and wind power plant can not be generated over a longer time because of bad weather. The sharp decline of expensive diesel imports means that electricity is cheaper than before. The so generated profit will each divided in half between the investor of the new plant and the end users. More Azores islands are to follow.

## 80 MWh of Storage in California

Tesla installed a grid storage facility for the Southern California Edison with a capacity of 80 MWh at a power of 20 MW between September 2016 and December 2016. This means that the storage unit (1/2017) is currently one of the largest accumulator batteries on the market. Tesla installed 400 Powerpack-2 modules at the Mira Loma transformer station in California. The memory serves to store energy at a low network load and then to feed this energy back into the grid at peak load. Prior to this, gas-fired power stations were used.

## 300 MWh Storage in Japan

Mitsubishi installed a storage facility in Buzen, Fukuoka Prefecture in Japan with 300 MWh capacity and 50 MW power. The storage is used to stabilize the network to compensate for fluctuations caused by renewable energies. The accumulator is in the power range of pumped storage power plants. The batteries are installed in 252 containers. The plant occupies an area of 14,000 square meters.

## Largest Grid Batteries

| Name | Commissioning date | Energy (MWh) | Power (MW) | Duration (hours) | Type | Country |
|------|------|------|------|------|------|------|
| Buzen Substation | 3 March 2016 | 300 | 50 | 6 | Sodium-sulphur | Japan |
| Escondido Substation | 24 February 2017 | 120 | 30 | 4 | Lithium-ion | USA |
| Pomona Substation | January 2017 | 80 | 20 | 4 | Lithium-ion | USA |
| Mira Loma Substation | 30 Jan. 2017 | 80 | 20 | 4 | Lithium-ion | USA |
| Minamisōma Substation | February 2016 | 40 | 40 | | Lithium-ion | Japan |

## Market Development

In the US, the market for storage power plants in 2015 has increased by 243 percent compared to, 2014. In 2016, the UK grid operator National Grid posted independent from technology 200 MW of control power to increase system stability. In this case, only battery storage power plants won the auction.

# Flywheel Storage Power System

A flywheel-storage power system uses a flywheel for energy storage, and can be a comparatively small storage facility with a peak power of up to 20 MW. It typically is used to stabilize to some degree power grids, to help them stay on the grid frequency, and to serve as a short-term compensation storage. Unlike common storage power plants, such as the pumped storage power plants with capacities up to 1000 MW, the benefits from flywheel storage power plants can be obtained with a facility in the range of a few kW to several 10 MW. They are comparable in this application with battery storage power plants.

Possible areas of application are places where electrical energy can be obtained and stored, and must be supplied again to compensate for example, fluctuations in the seconds range in wind or solar power. These storage facilities consist of individual flywheels in a modular design. Energy up to 150 kW can be absorbed or released per flywheel. Through combinations of several such flywheel accumulators, which are individually housed in buried underground vacuum tanks, a total power of up to several 10 MW can be achieved. The electrical connections power low voltage motors via a DC intermediate circuit and the power converter systems are comparable to those found in plants used in the high-voltage direct current transmissions application.

Sometimes battery storage power stations are built with flywheel storage power systems in order to conserve battery power. Flywheels can handle rapid fluctuations better.

## Application Examples

In vehicles small storage of power flywheels are used as an additional mechanism with batteries, to store the braking energy by regeneration. Power can be stored in the short term and then released back into the acceleration phase of a vehicle with very large electrical currents. This conserves battery power.

Flywheel storage for trams are a good application. During braking (such as when arriving at a station), high energy peaks are found which can not be always fed back into the power grid due to overloading danger. The flywheel energy storage power plants are in containers on side of the tracks and take the excess electrical energy. For example, up to 200 000 kWh energy per brake system are annually recovered in Zwickau.

In Stephentown, New York, Beacon Power operates in a flywheel storage power plant with 200 flywheels of 25 kWh capacity and 100 kW of power, Ganged together this gives 5 MWh capacity and 20 MW of power. The units operate at a peak speed at 15,000 rpm. The rotor flywheel consists of wound CFRP fibers which are filled with resin. The installation is intended primarily for frequency control. This service is sold to the New York power grid.

Stadtwerke München (SWM, Munich, Germany) uses a flywheel storage power system to stabilize the power grid, as well as control energy and to compensate for deviations from renewable energy sources. The plant originates from the Jülich Stornetic GmbH. The system consists of 28 flywheels and has a capacity of 100 kWh and a capacity of 600 kilovolt-amperes (kVA). The flywheels rotate at a peak speed of 45,000 rpm.

In Ontario, Canada, Temporal Power Ltd. has operated a flywheel storage power plant since 2014. It consists of 10 flywheels made of steel. Each flywheel weighs four tons and is 2.5 meters high. The maximum number of revolutions is at 11,500 revolutions per minute. The maximum power is 2 MW. The system is used for frequency regulation. After a successful three-year trial period, the system is to be expanded to 20MW and 100MW.

On the island of Aruba is currently a 5MW flywheel storage power plant built by Temporal Power Ltd. The island of Aruba intends to convert its energy supply to 100 percent renewables by 2020.

The city of Fresno in California is running flywheel storage power plants built by Amber Kinetics to store solar energy, which is produced in excess quantity in the daytime, for consumption at night.

## Energy Loss

It is now possible to build a flywheel storage system that loses just 5 percent of the energy stored in it, per day (i.e.the self-discharge rate).

## Flywheel Energy Storage

Flywheel energy storage (FES) works by accelerating a rotor (flywheel) to a very high speed and maintaining the energy in the system as rotational energy. When energy is extracted from the system, the flywheel's rotational speed is reduced as a consequence of the principle of conservation of energy; adding energy to the system correspondingly results in an increase in the speed of the flywheel.

Most FES systems use electricity to accelerate and decelerate the flywheel, but devices that directly use mechanical energy are being developed.

Advanced FES systems have rotors made of high strength carbon-fiber composites, suspended by magnetic bearings, and spinning at speeds from 20,000 to over 50,000 rpm in a vacuum enclosure. Such flywheels can come up to speed in a matter of minutes – reaching their energy capacity much more quickly than some other forms of storage.

## Main Components

A typical system consists of a rotor suspended by bearings inside a vacuum chamber to reduce friction, connected to a combination electric motor and electric generator.

First generation flywheel energy storage systems use a large steel flywheel rotating on mechanical bearings. Newer systems use carbon-fiber composite rotors that have a higher tensile strength than steel and are an order of magnitude less heavy.

Magnetic bearings are sometimes used instead of mechanical bearings, to reduce friction.

Other components are hub and shaft.

## Possible Future use of Superconducting Bearings

The expense of refrigeration led to the early dismissal of low-temperature superconductors for use in magnetic bearings. However, high-temperature superconductor (HTSC) bearings may be economical and could possibly extend the time energy could be stored economically. Hybrid bearing systems are most likely to see use first. High-temperature superconductor bearings have historically had problems providing the lifting forces necessary for the larger designs, but can easily provide a stabilizing force. Therefore, in hybrid bearings, permanent magnets support the load and high-temperature superconductors are used to stabilize it. The reason superconductors can work well stabilizing the load is because they are perfect diamagnets. If the rotor tries to drift off center, a restoring force due to flux pinning restores it. This is known as the magnetic stiffness of the bearing. Rotational axis vibration can occur due to low stiffness and damping, which are inherent problems of superconducting magnets, preventing the use of completely superconducting magnetic bearings for flywheel applications.

Since flux pinning is an important factor for providing the stabilizing and lifting force, the HTSC can be made much more easily for FES than for other uses. HTSC powders can be formed into arbitrary shapes so long as flux pinning is strong. An ongoing challenge that has to be overcome before superconductors can provide the full lifting force for an FES system is finding a way to suppress the decrease of levitation force and the gradual fall of rotor during operation caused by the flux creep of the superconducting material.

## Physical Characteristics

### General

Compared with other ways to store electricity, FES systems have long lifetimes (lasting decades with little or no maintenance; full-cycle lifetimes quoted for flywheels range from in excess of $10^5$, up to $10^7$, cycles of use), high specific energy (100–130 W·h/kg, or 360–500 kJ/kg), and large maximum power output. The energy efficiency (*ratio of energy out per energy in*) of flywheels, also known as round-trip efficiency, can be as high as 90%. Typical capacities range from 3 kWh to 133 kWh. Rapid charging of a system occurs in less than 15 minutes. The high specific energies often cited with flywheels can be a little misleading as commercial systems built have much lower specific energy, for example 11 W·h/kg, or 40 kJ/kg.

### Specific Energy

The maximum specific energy of a flywheel rotor is mainly dependent on two factors, the first being the rotor's geometry, and the second being the properties of the material being used. For single-material, isotropic rotors this relationship can be expressed as

$$\frac{E}{I} = K\left(\frac{\sigma}{\rho}\right),$$

where the variables are defined as follows:

$E$ - kinetic energy of the rotor [J]

$I$ - the rotor's moment of inertia [kg.m²]

$K$ - the rotor's geometric shape factor [m⁻²]

$\sigma$ - the tensile strength of the material [Pa]

$\rho$ - the material's density [kg/m³]

## Geometry (Shape Factor)

The highest possible value for the shape factor of a flywheel rotor, is $K = 1$, which can only be achieved by the theoretical *constant-stress disc* geometry. A constant-thickness disc geometry has a shape factor of $K = 0.606$, while for a rod of constant thickness the value is $K = 0.333$. A thin cylinder has a shape factor of $K = 0.5$.

## Material Properties

For energy storage purposes, materials with high strength, and low density are desirable. For this reason, composite materials are frequently being used in advanced flywheels. The strength-to-density ratio of a material can be expressed in Wh/kg (or Nm/kg); values greater than 400 Wh/kg can be achieved by certain composite materials.

## Rotor Materials

Several modern flywheel rotors are made from composite materials. Examples include the carbon fiber composite flywheel from Beacon Power Corporation and the *PowerThru* flywheel from Phillips Service Industries. Alternatively, Calnetix utilizes aerospace grade high-performance steel in their flywheel construction.

For these rotors, the relationship between material properties, geometry and energy density can be expressed by using a weighed-average approach.

## Tensile Strength and Failure Modes

One of the primary limits to flywheel design is the tensile strength of the material used for the rotor. Generally speaking, the stronger the disc, the faster it may be spun, and the more energy the system can store.

When the tensile strength of a composite flywheel's outer binding cover is exceeded, the binding cover will fracture, followed by the wheel shattering as the outer wheel compression is lost around the entire circumference, releasing all of its stored energy at once; this is commonly referred to as "flywheel explosion" since wheel fragments can reach kinetic energy comparable to that of a bullet. Composite materials that are wound and glued in layers tend to disintegrate quickly, first into small-diameter filaments that entangle and slow each other, and then into red-hot powder, instead of large chunks of high-velocity shrapnel as can occur with a cast metal flywheel.

For a cast metal flywheel, the failure limit is the binding strength of the grain boundaries of the polycrystalline molded metal. Aluminum in particular suffers from fatigue and can develop microfractures due to repeated low-energy stretching. Angular forces may cause portions of a metal flywheel to bend outward and begin dragging on the outer containment vessel, or to separate completely and bounce randomly around the interior. The rest of the flywheel is now severely unbalanced, which may lead to rapid bearing failure from vibration, and sudden shock fracturing of large segments of the flywheel.

Traditional flywheel systems require strong containment vessels as a safety precaution, which increases the total mass of the device. The energy release from failure can be dampened with a gelatinous or encapsulated liquid inner housing lining, which will boil and absorb the energy of destruction. Still, many customers of large-scale flywheel energy-storage systems prefer to have them embedded in the ground to halt any material that might escape the containment vessel.

## Energy Storage Efficiency

Flywheel energy storage systems using mechanical bearings can lose 20% to 50% of their energy in two hours. Much of the friction responsible for this energy loss results from the flywheel changing orientation due to the rotation of the earth (an effect similar to that shown by a Foucault pendulum). This change in orientation is resisted by the gyroscopic forces exerted by the flywheel's angular momentum, thus exerting a force against the mechanical bearings. This force increases friction. This can be avoided by aligning the flywheel's axis of rotation parallel to that of the earth's axis of rotation.

Conversely, flywheels with magnetic bearings and high vacuum can maintain 97% mechanical efficiency, and 85% round trip efficiency.

## Effects of Angular Momentum in Vehicles

When used in vehicles, flywheels also act as gyroscopes, since their angular momentum is typically of a similar order of magnitude as the forces acting on the moving vehicle. This property may be detrimental to the vehicle's handling characteristics while turning or driving on rough ground; driving onto the side of a sloped embankment may cause wheels to partially lift off the ground as the flywheel opposes sideways tilting forces. On the other hand, this property could be utilized to keep the car balanced so as to keep it from rolling over during sharp turns.

When a flywheel is used entirely for its effects on the attitude of a vehicle, rather than for energy storage, it is called a reaction wheel or a control moment gyroscope.

The resistance of angular tilting can be almost completely removed by mounting the flywheel within an appropriately applied set of gimbals, allowing the flywheel to retain its original orientation without affecting the vehicle. This doesn't avoid the complication of gimbal lock, and so a compromise between the number of gimbals and the angular freedom is needed.

The center axle of the flywheel acts as a single gimbal, and if aligned vertically, allows for the 360 degrees of yaw in a horizontal plane. However, for instance driving up-hill requires a second pitch gimbal, and driving on the side of a sloped embankment requires a third roll gimbal.

## Full-motion Gimbals

Although the flywheel itself may be of a flat ring shape, a free-movement gimbal mounting inside a vehicle requires a spherical volume for the flywheel to freely rotate within. Left to its own, a spinning flywheel in a vehicle would slowly precess following the Earth's rotation, and precess further yet in vehicles that travel long distances over the Earth's curved spherical surface.

A full-motion gimbal has additional problems of how to communicate power into and out of the flywheel, since the flywheel could potentially flip completely over once a day, precessing as the Earth rotates. Full free rotation would require slip rings around each gimbal axis for power conductors, further adding to the design complexity.

## Limited-motion Gimbals

To reduce space usage, the gimbal system may be of a limited-movement design, using shock absorbers to cushion sudden rapid motions within a certain number of degrees of out-of-plane angular rotation, and then gradually forcing the flywheel to adopt the vehicle's current orientation. This reduces the gimbal movement space around a ring-shaped flywheel from a full sphere, to a short thickened cylinder, encompassing for example ± 30 degrees of pitch and ± 30 degrees of roll in all directions around the flywheel.

## Counterbalancing of Angular Momentum

An alternative solution to the problem is to have two joined flywheels spinning synchronously in opposite directions. They would have a total angular momentum of zero and no gyroscopic effect. A problem with this solution is that when the difference between the momentum of each flywheel is anything other than zero the housing of the two flywheels would exhibit torque. Both wheels must be maintained at the same speed to keep the angular velocity at zero. Strictly speaking, the two flywheels would exert a huge torqueing moment at the central point, trying to bend the axle. However, if the axle were sufficiently strong, no gyroscopic forces would have a net effect on the sealed container, so no torque would be noticed.

To further balance the forces and spread out strain, a single large flywheel can be balanced by two half-size flywheels on each side, or the flywheels can be reduced in size to be a series of alternating layers spinning in opposite directions. However this increases housing and bearing complexity.

## Applications

### Automotive

In the 1950s, flywheel-powered buses, known as gyrobuses, were used in Yverdon (Switzerland) and Ghent (Belgium) and there is ongoing research to make flywheel systems that are smaller, lighter, cheaper and have a greater capacity. It is hoped that flywheel systems can replace conventional chemical batteries for mobile applications, such as for electric vehicles. Proposed flywheel systems would eliminate many of the disadvantages of existing battery power systems, such as low capacity, long charge times, heavy weight and short usable lifetimes. Flywheels may have been used in the experimental Chrysler Patriot, though that has been disputed.

Flywheels have also been proposed for use in continuously variable transmissions. Punch Powertrain is currently working on such a device.

During the 1990s, Rosen Motors developed a gas turbine powered series hybrid automotive powertrain using a 55,000 rpm flywheel to provide bursts of acceleration which the small gas turbine engine could not provide. The flywheel also stored energy through regenerative braking. The flywheel was composed of a titanium hub with a carbon fiber cylinder and was gimbal-mounted to minimize adverse gyroscopic effects on vehicle handling. The prototype vehicle was successfully road tested in 1997 but was never mass-produced.

In 2013, Volvo announced a flywheel system fitted to the rear axle of its S60 sedan. Braking action spins the flywheel at up to 60,000 rpm and stops the front-mounted engine. Flywheel energy is applied via a special transmission to partially or completely power the vehicle. The 20-centimetre (7.9 in), 6-kilogram (13 lb) carbon fiber flywheel spins in a vacuum to eliminate friction. When partnered with a four-cylinder engine, it offers up to a 25 percent reduction in fuel consumption versus a comparably performing turbo six-cylinder, providing an 80 hp boost and allowing it to reach 100 kilometres per hour (62 mph) in 5.5 seconds. The company did not announce specific plans to include the technology in its product line.

In July 2014 GKN acquired Williams Hybrid Power (WHP) division and intends to supply 500 carbon fiber *Gyrodrive* electric flywheel systems to urban bus operators over the next two years As the former developer name implies, these were originally designed for Formula one motor racing applications. In September 2014, Oxford Bus Company announced that it is introducing 14 *Gyrodrive hybrid* buses by Alexander Dennis on its Brookes Bus operation.

## Rail Vehicles

Flywheel systems have been used experimentally in small electric locomotives for shunting or switching, e.g. the Sentinel-Oerlikon Gyro Locomotive. Larger electric locomotives, e.g. British Rail Class 70, have sometimes been fitted with flywheel boosters to carry them over gaps in the third rail. Advanced flywheels, such as the 133 kWh pack of the University of Texas at Austin, can take a train from a standing start up to cruising speed.

The Parry People Mover is a railcar which is powered by a flywheel. It was trialled on Sundays for 12 months on the Stourbridge Town Branch Line in the West Midlands, England during 2006 and 2007 and was intended to be introduced as a full service by the train operator London Midland in December 2008 once two units had been ordered. In January 2010, both units are in operation.

## Rail Electrification

FES can be used at the lineside of electrified railways to help regulate the line voltage thus improving the acceleration of unmodified electric trains and the amount of energy recovered back to the line during regenerative braking, thus lowering energy bills. Trials have taken place in London, New York, Lyon and Tokyo, and New York MTA's Long Island Rail Road is now investing $5.2m in a pilot project on LIRR's West Hempstead Branch line. These trials and systems store kinetic energy in rotors consisting of a carbon-glass composite cylinder packed with neodymium-iron-boron powder that forms a permanent magnet. These spin at up to 37800rev/min, and each 100 kW unit can store 11 MJ of re-usable energy.

## Uninterruptible Power Supplies

Flywheel power storage systems in production as of 2001 have storage capacities comparable to batteries and faster discharge rates. They are mainly used to provide load leveling for large battery systems, such as an uninterruptible power supply for data centers as they save a considerable amount of space compared to battery systems.

Flywheel maintenance in general runs about one-half the cost of traditional battery UPS systems. The only maintenance is a basic annual preventive maintenance routine and replacing the bearings every five to ten years, which takes about four hours. Newer flywheel systems completely levitate the spinning mass using maintenance-free magnetic bearings, thus eliminating mechanical bearing maintenance and failures.

Costs of a fully installed flywheel UPS (including power conditioning) are (in 2009) about $330 per kilowatt (for 15 seconds full-load capacity).

## Test Laboratories

A long-standing niche market for flywheel power systems are facilities where circuit-breakers and similar devices are tested: even a small household circuit-breaker may be rated to interrupt a current of 10,000 or more amperes, and larger units may have interrupting ratings of 100,000 or 1,000,000 amperes. The enormous transient loads produced by deliberately forcing such devices to demonstrate their ability to interrupt simulated short circuits would have unacceptable effects on the local grid if these tests were done directly from building power. Typically such a laboratory will have several large motor-generator sets, which can be spun up to speed over some minutes; then the motor is disconnected before a circuit breaker is tested.

## Physics Laboratories

Tokamak fusion experiments need very high currents for brief intervals (mainly to power large electromagnets for a few seconds).

- JET (the Joint European Torus) has two 775 ton flywheels (installed in 1981) that spin up to 225 rpm. Each flywheel stores 3.75 GJ and can deliver at up to 400MW.

- ASDEX has 3 flywheel generators.

- DIII-D tokamak at General Atomics

- the Princeton Large Torus (PLT) at the Princeton Plasma Physics Laboratory

Also the non-tokamak: Nimrod synchrotron at the Rutherford Appleton Laboratory had two 30 ton flywheels.

## Aircraft Launching Systems

The *Gerald R. Ford*-class aircraft carrier will use flywheels to accumulate energy from the ship's power supply, for rapid release into the electromagnetic aircraft launch system. The shipboard power system cannot on its own supply the high power transients necessary to launch aircraft. Each of four rotors will store 121 MJ at 6400 rpm. They can store 122 MJ in 45 secs and release it in 2–3 seconds. The flywheel energy densities are 28 kJ/kg; including the stators and cases this comes down to 18.1 kJ/kg, excluding the torque frame.

## NASA G2 Flywheel for Spacecraft Energy Storage

This was a design funded by NASA's Glenn Research Center and intended for component testing in a laboratory environment. It used a carbon fiber rim with a titanium hub designed to spin at 60,000 rpm, mounted on magnetic bearings. Weight was limited to 250 pounds. Storage was 525 W-hr (1.89 MJ) and could be charged or discharged at 1 kW. The working model shown in the photograph at the top of the page ran at 41,000 rpm on September 2, 2004.

## Amusement Rides

The Incredible Hulk roller coaster at Universal's Islands of Adventure features a rapidly accelerating uphill launch as opposed to the typical gravity drop. This is achieved through powerful traction motors that throw the car up the track. To achieve the brief very high current required to accelerate a full coaster train to full speed uphill, the park utilizes several motor generator sets with large flywheels. Without these stored energy units, the park would have to invest in a new substation or risk browning-out the local energy grid every time the ride launches.

## Pulse Power

Compulsators (low-inductance alternators) act like capacitors, they can be can be spun up to provide pulsed power for railguns and lasers. Instead of having a separate flywheel and generator, only the large rotor of the alternator stores energy.

## Motor Sports

Using a continuously variable transmission (CVT), energy is recovered from the drive train during braking and stored in a flywheel. This stored energy is then used during acceleration by altering the ratio of the CVT. In motor sports applications this energy is used to improve acceleration rather than reduce carbon dioxide emissions – although the same technology can be applied to road cars to improve fuel efficiency.

Automobile Club de l'Ouest, the organizer behind the annual 24 Hours of Le Mans event and the Le Mans Series, is currently "studying specific rules for LMP1 which will be equipped with a kinetic energy recovery system."

Williams Hybrid Power, a subsidiary of Williams F1 Racing team, have supplied Porsche and Audi with flywheel based hybrid system for Porsche's 911 GT3 R Hybrid and Audi's R18 e-Tron Quattro. Audi's victory in 2012 24 Hours of Le Mans is the first for a hybrid(diesel-electric) vehicle.

## Grid Energy Storage

Flywheels are sometimes used as short term spinning reserve for momentary grid frequency regulation and balancing sudden changes between supply and consumption. No carbon emissions, faster response times and ability to buy power at off-peak hours are among the advantages of using flywheels instead of traditional sources of energy like natural gas turbines. Operation is very similar to batteries in the same application, their differences are primarily economic.

Beacon Power opened a 5 MWh (20 MW over 15 mins) flywheel energy storage plant in Stephentown, New York in 2011 using 200 flywheels and a similar 20 MW system at Hazle Township, Pennsylvania in 2014.

A 2 MW (for 15 min) flywheel storage facility in Minto, Ontario, Canada opened in 2014. The flywheel system (developed by NRStor) uses 10 spinning steel flywheels on magnetic bearings.

Amber Kinetics, Inc. of Union City, CA has an agreement with Pacific Gas and Electric (PG&E) for a 20 MW / 80 MWh flywheel energy storage facility located in Fresno, CA, the first utility-scale flywheel project to feature game changing, four-hour discharge duration.

## Wind Turbines

Flywheels may be used to store energy generated by wind turbines during off-peak periods or during high wind speeds.

Beacon Power began testing of their Smart Energy 25 (Gen 4) flywheel energy storage system at a wind farm in Tehachapi, California. The system is part of a wind power/flywheel demonstration project being carried out for the California Energy Commission.

## Toys

Friction motors used to power many toy cars, trucks, trains, action toys and such, are simple flywheel motors.

## Toggle Action Presses

In industry, toggle action presses are still popular. The usual arrangement involves a very strong crankshaft and a heavy duty connecting rod which drives the press. Large and heavy flywheels are driven by electric motors but the flywheels only turn the crankshaft when clutches are activated.

## Comparison to Electric Batteries

Flywheels are not as adversely affected by temperature changes, can operate at a much wider temperature range, and are not subject to many of the common failures of chemical rechargeable batteries. They are also less potentially damaging to the environment, being largely made of inert or benign materials. Another advantage of flywheels is that by a simple measurement of the rotation speed it is possible to know the exact amount of energy stored.

Unlike most batteries which only operate for a finite period (for example roughly 36 months in the case of lithium ion polymer batteries), a flywheel potentially has an indefinite working lifespan. Flywheels built as part of James Watt steam engines have been continuously working for more than two hundred years. Working examples of ancient flywheels used mainly in milling and pottery can be found in many locations in Africa, Asia, and Europe.

Most modern flywheels are typically a sealed device that needs minimal maintenance throughout its service life. Magnetic bearing flywheels in a vacuum enclosure, such as the NASA model depicted above, do not need any bearing maintenance and are therefore superior to batteries both in terms of total lifetime and energy storage capacity. Flywheel systems with mechanical bearings will have a limited lifespan due to wear.

The arrangement of batteries can be designed to a wide variety of configurations, whereas a flywheel at a minimum must occupy a square surface area. Where space is a constraint for the application of energy storage (e.g. under trains in tunnels) the flywheel may not be a valid application.

## References

- "A cheap, long-lasting, sustainable battery for grid energy storage | KurzweilAI". www.kurzweilai.net. 2016-09-16. Retrieved 2017-02-02

- Castelvecchi, Davide (May 19, 2007). "Spinning into control: High-tech reincarnations of an ancient way of storing energy". Science News. 171 (20): 312–313. doi:10.1002/scin.2007.5591712010

- MICU, ALEXANDRU (2017-01-30). "Rows of Tesla batteries will keep Southern California's lights on during the night". ZME Science. Retrieved 2017-02-02

- Lucien F. Trueb; Paul Rüetschi (1998) (in German), Batterien und Akkumulatoren, Springer, pp. 85 bis 89, ISBN 3-540-62997-1

- Genta, Giancarlo (1989). "Some considerations on the constant stress disc profile". Meccanica. 24 (4): 235–248. doi:10.1007/BF01556455

- Chambers's Encyclopaedia: A Dictionary of Universal Knowledge. W. & R. Chambers, LTD. 1896. pp. 252–253. Retrieved January 7, 2009

- Donners, K.; Waelkens, M.; Deckers, J. (2002). "Water Mills in the Area of Sagalassos: A Disappearing Ancient Technology". Anatolian Studies. 52: 1–17. JSTOR 3643076. doi:10.2307/3643076

- Wakefield, Ernest (1998). History of the Electric Automobile: Hybrid Electric Vehicles. SAE. p. 332. ISBN 0-7680-0125-0

- "A look at the new battery storage facility in California built with Tesla Powerpacks". Ars Technica. 30 January 2017. Retrieved 6 February 2017

# Hydrogen, Thermal Energy and Oil Storage

Hydrogen storage for later use is usually done by converting hydrogen into liquid hydrogen and slush hydrogen. Liquid hydrogen is stored using cryogenic processes. Compressed hydrogen is stored in tankers and large storage units. The topics discussed in the chapter are of great importance to broaden the existing knowledge on energy storage.

## Hydrogen Storage

Methods of hydrogen storage for subsequent use span many approaches including high pressures, cryogenics, and chemical compounds that reversibly release $H_2$ upon heating. Underground hydrogen storage is useful to provide grid energy storage for intermittent energy sources, like wind power, as well as providing fuel for transportation, particularly for ships and airplanes.

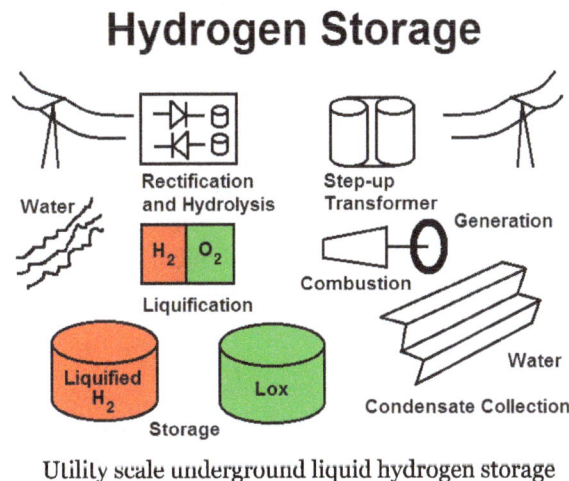

Utility scale underground liquid hydrogen storage

Most research into hydrogen storage is focused on storing hydrogen as a lightweight, compact energy carrier for mobile applications.

Liquid hydrogen or slush hydrogen may be used, as in the Space Shuttle. However liquid hydrogen requires cryogenic storage and boils around 20.268 K (−252.882 °C or −423.188 °F). Hence, its liquefaction imposes a large energy loss (as energy is needed to cool it down to that temperature). The tanks must also be well insulated to prevent boil off but adding insulation increases cost. Liquid hydrogen has less energy density *by volume* than hydrocarbon fuels such as gasoline by approximately a factor of four. This highlights the density problem for pure hydrogen: there is actually about 64% more hydrogen in a liter of gasoline (116 grams hydrogen) than there is in a liter of pure liquid hydrogen (71 grams hydrogen). The carbon in the gasoline also contributes to the energy of combustion.

Compressed hydrogen, by comparison, is stored quite differently. Hydrogen gas has good energy density by weight, but poor energy density by volume versus hydrocarbons, hence it requires a larger tank to store. A large hydrogen tank will be heavier than the small hydrocarbon tank used to store the same amount of energy, all other factors remaining equal. Increasing gas pressure would improve the energy density by volume, making for smaller, but not lighter container tanks. Compressed hydrogen costs 2.1% of the energy content to power the compressor. Higher compression without energy recovery will mean more energy lost to the compression step. Compressed hydrogen storage can exhibit very low permeation.

## Established Technologies

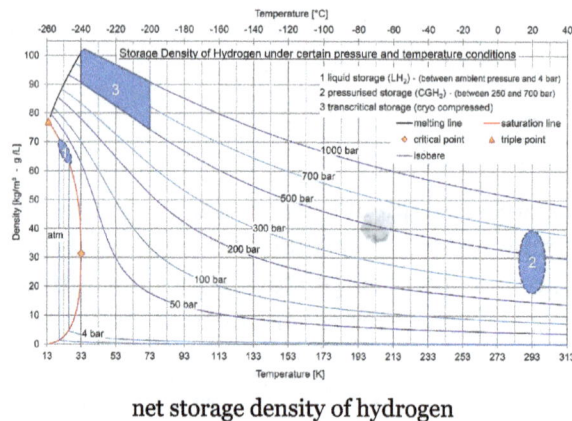

net storage density of hydrogen

## Compressed Hydrogen

Compressed hydrogen is a storage form where hydrogen gas is kept under pressures to increase the storage density. Compressed hydrogen in hydrogen tanks at 350 bar (5,000 psi) and 700 bar (10,000 psi) is used for hydrogen tank systems in vehicles, based on type IV carbon-composite technology. Car manufacturers have been developing this solution, such as Honda or Nissan.

## Liquid Hydrogen

BMW has been working on liquid hydrogen tanks for cars, producing for example the BMW Hydrogen 7.

## Proposals and Research

Hydrogen storage technologies can be divided into physical storage, where hydrogen molecules are stored (including pure hydrogen storage via compression and liquefaction), and chemical storage, where hydrides are stored.

## Chemical Storage

Chemical storage could offer high storage performance due to the strong binding of hydrogen and the high storage densities. However, the regeneration of storage material is still an issue. A large number of chemical storage systems are under investigation, which involve hydrolysis reactions, hydrogenation/dehydrogenation reactions, ammonia borane and other boron hydrides, ammonia, and alane etc. Storage in hydrocarbons may also be successful in overcoming the issue with low density. For example, supercritical hydrogen at 30 °C and 500 bar only has a density of 15.0 mol/L while methanol has a density of 49.5 mol $H_2$/L methanol and saturated dimethyl ether at 30 °C and 7 bar has a density of 42.1 mol $H_2$/L dimethyl ether. These liquids would use much smaller, cheaper, safer storage tanks.

## Metal Hydrides

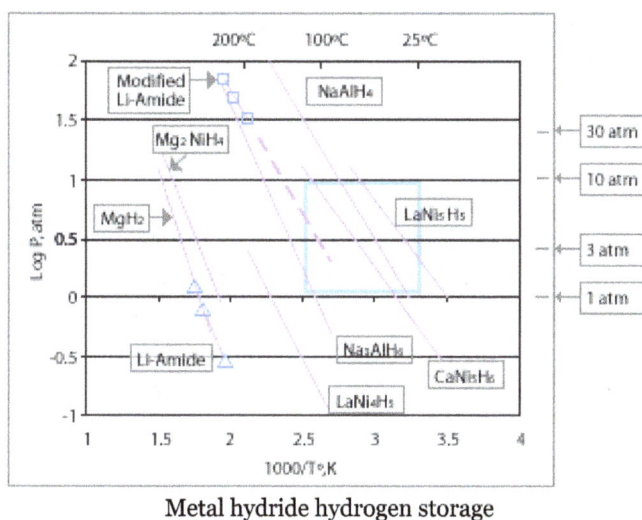

Metal hydride hydrogen storage

Metal hydrides, such as $MgH_2$, $NaAlH_4$, $LiAlH_4$, LiH, $LaNi_5H_6$, $TiFeH_2$ and palladium hydride, with varying degrees of efficiency, can be used as a storage medium for hydrogen, often reversibly. Some are easy-to-fuel liquids at ambient temperature and pressure, others are solids which could be turned into pellets. These materials have good energy density by volume, although their energy density by weight is often worse than the leading hydrocarbon fuels.

Most metal hydrides bind with hydrogen very strongly. As a result, high temperatures around 120 °C (248 °F) – 200 °C (392 °F) are required to release their hydrogen content. This energy cost can be reduced by using alloys which consists of a strong hydride former and a weak one such as in $LiNH_2$, $LiBH_4$ and $NaBH_4$. These are able to form weaker bonds, thereby requiring less input to release stored hydrogen. However, if the interaction is too weak, the pressure needed for rehydriding is high, thereby eliminating any energy savings. The target for onboard hydrogen fuel systems is roughly <100 °C for release and <700 bar for recharge (20–60 kJ/mol $H_2$).

An alternative method for reducing dissociation temperatures is doping with activators. This has been successfully used for aluminium hydride but its complex synthesis makes it undesirable for most applications as it is not easily recharged with hydrogen.

Currently the only hydrides which are capable of achieving the 9 wt% gravimetric goal for 2015 are limited to lithium, boron and aluminium based compounds; at least one of the first-row elements or Al must be added. Research is being done to determine new compounds which can be used to meet these requirements.

Proposed hydrides for use in a hydrogen economy include simple hydrides of magnesium or transition metals and complex metal hydrides, typically containing sodium, lithium, or calcium and aluminium or boron. Hydrides chosen for storage applications provide low reactivity (high safety) and high hydrogen storage densities. Leading candidates are lithium hydride, sodium borohydride, lithium aluminium hydride and ammonia borane. A French company McPhy Energy is developing the first industrial product, based on magnesium hydride, already sold to some major clients such as Iwatani and ENEL.

*New Scientist* reported that Arizona State University is investigating using a borohydride solution to store hydrogen, which is released when the solution flows over a catalyst made of ruthenium. Researchers at University of Pittsburgh and Georgia Tech performed extensive benchmarking simulations on mixtures of several light metal hydrides to predict possible reaction thermodynamics for hydrogen storage.

## Non-metal Hydrides

The Italian catalyst manufacturer Acta has proposed using hydrazine as an alternative to hydrogen in fuel cells. As the hydrazine fuel is liquid at room temperature, it can be handled and stored more easily than hydrogen. By storing it in a tank full of a double-bonded carbon-oxygen carbonyl, it reacts and forms a safe solid called hydrazone. By then flushing the tank with warm water, the liquid hydrazine hydrate is released. Hydrazine breaks down in the cell to form nitrogen and hydrogen which bonds with oxygen, releasing water.

## Carbohydrates

Carbohydrates (polymeric $C_6H_{10}O_5$) releases $H_2$ in a bioreformer mediated by the enzyme cocktail—cell-free synthetic pathway biotransformation. Carbohydrate provides high hydrogen storage densities as a liquid with mild pressurization and cryogenic constraints: It can also be stored as a solid powder. Carbohydrate is the most abundant renewable bioresource in the world.

In May 2007 biochemical engineers from the Virginia Polytechnic Institute and State University and biologists and chemists from the Oak Ridge National Laboratory announced a method of producing high yield pure hydrogen from starch and water. In 2009, they demonstrated to produce nearly 12 moles of hydrogen per glucose unit from cellulosic materials and water. Thanks to complete conversion and modest reaction conditions, they propose to use carbohydrate as a high energy density hydrogen carrier with a density of 14.8 wt%.

## Synthesized Hydrocarbons

An alternative to hydrides is to use regular hydrocarbon fuels as the hydrogen carrier. Then a small hydrogen reformer would extract the hydrogen as needed by the fuel cell. However, these reformers are slow to react to changes in demand and add a large incremental cost to the vehicle powertrain.

Direct methanol fuel cells do not require a reformer, but provide a lower energy density compared to conventional fuel cells, although this could be counterbalanced with the much better energy densities of ethanol and methanol over hydrogen. Alcohol fuel is a renewable resource.

Solid-oxide fuel cells can operate on light hydrocarbons such as propane and methane without a reformer, or can run on higher hydrocarbons with only partial reforming, but the high temperature and slow startup time of these fuel cells are problematic for automotive applications.

## Liquid Organic Hydrogen Carriers (LOHC)

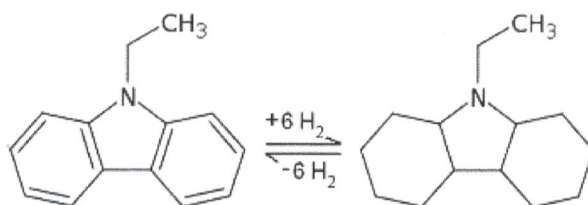

Unsaturated organic compounds can store huge amounts of hydrogen. These *Liquid Organic Hydrogen Carriers* (LOHC) are hydrogenated for storage and dehydrogenated again when the energy/hydrogen is needed. Research on LOHC was concentrated on cycloalkanes at an early stage, with its relatively high hydrogen capacity (6-8 wt %) and production of $CO_x$-free hydrogen. Heterocyclic aromatic compounds (or N-Heterocycles) are also appropriate for this task. A compound that stands in the focus of the current LOHC research is N-ethylcarbazole (NEC) but many others do exist. More recently dibenzyltoluene, which is already industrially used as a heat transfer fluid in industry, was identified as potential LOHC. With a wide liquid range between -39 °C (melting point) and 390 °C (boiling point) and a hydrogen storage density of 6.2 wt% dibenzyltoluene is ideally suited as LOHC material. More recently, formic acid (FA) has been suggested as a promising hydrogen storage material with a 4.4wt% hydrogen capacity.

Using LOHCs relatively high gravimetric storage densities can be reached (about 6 wt-%) and the overall energy efficiency is higher than for other chemical storage options such as producing methane from the hydrogen.

- Cycloalkanes

Cycloalkanes reported as LOHC include cyclohexane, methyl-cyclohexane and decalin. The dehydrogenation of cycloalkanes is highly endothermic (63-69 kJ/mol $H_2$), which means this process requires high temperature. Dehydrogenation of decalin is the most thermodynamically favored among the three cycloalkanes, and methyl-cyclohexane is second because of the presence of the methyl group. Research on catalyst development for dehydrogenation of cycloalkanes has been carried out for decades. Nickel (Ni), Molybdenum (Mo) and Platinum (Pt) based catalysts are highly investigated for dehydrogenation. However, coking is still a big challenge for catalyst's long-term stability.

- N-Heterocycles

Both hydrogenation and dehydrogenation of LOHCs requires catalysts. It was demonstrated that replacing hydrocarbons by hetero-atoms, like N, O etc. improves reversible de/hydrogenation properties. The temperature required for hydrogenation and dehydrogenation of drops significantly with increasing numbers of heteroatoms. Among all the N-heterocycles, the saturated-unsaturated pair of dodecahydro-N-ethylcarbazole (12H-NEC) and NEC has been considered as a promising candidate for hydrogen storage with a fairly large hydrogen content (5.8wt%). The figure on the top right shows dehydrogenation and hydrogenation of the 12H-NEC and NEC pair. The standard catalyst for NEC to 12H-NEC is Ru and Rh based. The selectivity of hydrogenation can reach 97% at 7 MPa and 130 °C-150 °C. Although N-Heterocyles can optimize the unfavorable thermodynamic properties of cycloalkanes, a lot of issues remain unsolved, such as high cost, high toxicity and kinetic barriers etc.

- Formic acid

In 2006 researchers of EPFL, Switzerland, reported the use of formic acid as a hydrogen storage material. Carbon monoxide free hydrogen has been generated in a very wide pressure range (1–600 bar). A homogeneous catalytic system based on water-soluble ruthenium catalysts selectively decompose HCOOH into $H_2$ and $CO_2$ in aqueous solution. This catalytic system overcomes the limitations of other catalysts (e.g. poor stability, limited catalytic lifetimes, formation of CO) for the decomposition of formic acid making it a viable hydrogen storage material. And the co-product of this decomposition, carbon dioxide, can be used as hydrogen vector by hydrogenating it back to formic acid in a second step. The catalytic hydrogenation of $CO_2$ has long been studied and efficient procedures have been developed. Formic acid contains 53 g $L^{-1}$ hydrogen at room temperature and atmospheric pressure. By weight, pure formic acid stores 4.3 wt% hydrogen. Pure formic acid is a liquid with a flash point 69 °C (cf. gasoline –40 °C, ethanol 13 °C). 85% formic acid is not flammable.

## Ammonia

Ammonia ($NH_3$) releases $H_2$ in an appropriate catalytic reformer. Ammonia provides high hydrogen storage densities as a liquid with mild pressurization and cryogenic constraints: It can also be stored as a liquid at room temperature and pressure when mixed with water. Ammonia is the second most commonly produced chemical in the world and a large infrastructure for making, transporting, and distributing ammonia exists. Ammonia can be reformed to produce hydrogen with no harmful waste, or can mix with existing fuels and under the right conditions burn efficiently. Since there is no carbon in ammonia, no carbon by-products are produced; thereby making this possibility a "carbon neutral" option for the future. Pure ammonia burns poorly at the atmospheric pressures found in natural gas fired water heaters and stoves. Under compression in an automobile engine it is a suitable fuel for slightly modified gasoline engines. Ammonia is a toxic gas at normal temperature and pressure and has a potent odor.

In September 2005 chemists from the Technical University of Denmark announced a method of storing hydrogen in the form of ammonia saturated into a salt tablet. They claim it will be an inexpensive and safe storage method.

## Amine Borane Complexes

Prior to 1980, several compounds were investigated for hydrogen storage including complex borohydrides, or aluminohydrides, and ammonium salts. These hydrides have an upper theoretical hydrogen yield limited to about 8.5% by weight. Amongst the compounds that contain only B, N, and H (both positive and negative ions), representative examples include: amine boranes, boron hydride ammoniates, hydrazine-borane complexes, and ammonium octahydrotriborates or tetrahydroborates. Of these, amine boranes (and especially ammonia borane) have been extensively investigated as hydrogen carriers. During the 1970s and 1980s, the U.S. Army and Navy funded efforts aimed at developing hydrogen/deuterium gas-generating compounds for use in the HF/DF and HCl chemical lasers, and gas dynamic lasers. Earlier hydrogen gas-generating formulations used amine boranes and their derivatives. Ignition of the amine borane(s) forms boron nitride (BN) and hydrogen gas. In addition to ammonia borane ($H_3BNH_3$), other gas-generators include diborane diammoniate, $H_2B(NH_3)_2BH_4$.

## Imidazolium Ionic Liquids

In 2007 Dupont and others reported hydrogen-storage materials based on imidazolium ionic liquids. Simple alkyl(aryl)-3-methylimidazolium N-bis(trifluoromethanesulfonyl)imidate salts that possess very low vapour pressure, high density, and thermal stability and are not inflammable can add reversibly 6–12 hydrogen atoms in the presence of classical Pd/C or Ir0 nanoparticle catalysts and can be used as alternative materials for on-board hydrogen-storage devices. These salts can hold up to 30 g L$^{-1}$ of hydrogen at atmospheric pressure.

## Phosphonium Borate

In 2006 researchers of University of Windsor reported on reversible hydrogen storage in a non-metal phosphonium borate frustrated Lewis pair:

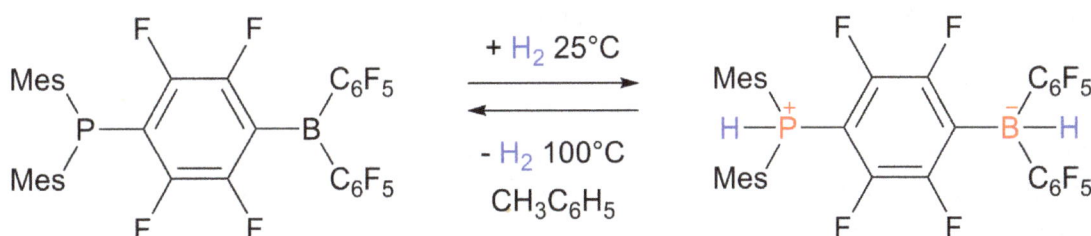

The phosphino-borane on the left accepts one equivalent of hydrogen at one atmosphere and 25 °C and expels it again by heating to 100 °C. The storage capacity is 0.25 wt% still rather below the 6 to 9 wt% required for practical use.

## Carbonite Substances

Research has proven that graphene can store hydrogen efficiently. After taking up hydrogen, the substance becomes graphane. After tests, conducted by dr André Geim at the University of Manchester, it was shown that not only can graphene store hydrogen easily, it can also release the hydrogen again, after heating to 450 °C.

## Metal-organic Frameworks

Metal-organic frameworks represent another class of synthetic porous materials that store hydrogen and energy at the molecular level. MOFs are highly crystalline inorganic-organic hybrid structures that contain metal clusters or ions (secondary building units) as nodes and organic ligands as linkers. When guest molecules (solvent) occupying the pores are removed during solvent exchange and heating under vacuum, porous structure of MOFs can be achieved without destabilizing the frame and hydrogen molecules will be adsorbed onto the surface of the pores by physisorption. Compared to traditional zeolites and porous carbon materials, MOFs have very high number of pores and surface area which allow higher hydrogen uptake in a given volume. Thus, research interests on hydrogen storage in MOFs have been growing since 2003 when the first MOF-based hydrogen storage was introduced. Since there are infinite geometric and chemical variations of MOFs based on different combinations of SBUs and linkers, many researches explore what combination will provide the maximum hydrogen uptake by varying materials of metal ions and linkers.

In 2006, chemists at UCLA and the University of Michigan have achieved hydrogen storage concentrations of up to 7.5 wt% in MOF-74 at a low temperature of 77 K. In 2009, researchers at University of Nottingham reached 10 wt% at 77 bar (1,117 psi) and 77 K with MOF NOTT-112. Most articles about hydrogen storage in MOFs report hydrogen uptake capacity at a temperature of 77K and a pressure of 1 bar because these conditions are commonly available and the binding energy between hydrogen and the MOF at this temperature is large compared to the thermal vibration energy. Varying several factors such as surface area, pore size, catenation, ligand structure, and sample purity can result in different amounts of hydrogen uptake in MOFs.

## Encapsulation

Cella Energy technology is based around the encapsulation of hydrogen gas and nano-structuring of chemical hydrides in small plastic balls, at room temperature and pressure.

## Physical Storage

In this case hydrogen remains in physical forms, i.e., as gas, supercritical fluid, adsorbate, or molecular inclusions. Theoretical limitations and experimental results are considered concerning the volumetric and gravimetric capacity of glass microvessels, microporous, and nanoporous media, as well as safety and refilling-time demands.

## Cryo-compressed

Cryo-compressed storage of hydrogen is the only technology that meets 2015 DOE targets for volumetric and gravimetric efficiency.

Furthermore, another study has shown that cryo-compressed exhibits interesting cost advantages: ownership cost (price per mile) and storage system cost (price per vehicle) are actually the lowest when compared to any other technology. For example, a cryo-compressed hydrogen system would cost $0.12 per mile (including cost of fuel and every associated other cost), while conventional gasoline vehicles cost between $0.05 and $0.07 per mile.

Like liquid storage, cryo-compressed uses cold hydrogen (20.3 K and slightly above) in order to reach a high energy density. However, the main difference is that, when the hydrogen would warm-up due to heat transfer with the environment ("boil off"), the tank is allowed to go to pressures much higher (up to 350 bars versus a couple of bars for liquid storage). As a consequence, it takes more time before the hydrogen has to vent, and in most driving situations, enough hydrogen is used by the car to keep the pressure well below the venting limit.

Consequently, it has been demonstrated that a high driving range could be achieved with a cryo-compressed tank : more than 650 miles (1,050 km) were driven with a full tank mounted on an hydrogen-fueled engine of Toyota Prius. Research is still on its way in order to study and demonstrate the full potential of the technology.

As of 2010, the BMW Group has started a thorough component and system level validation of cryo-compressed vehicle storage on its way to a commercial product.

# Carbon Nanotubes

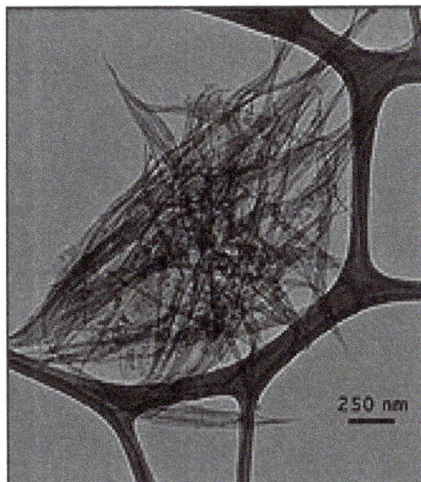

Carbon nanotubes

Hydrogen carriers based on nanostructured carbon (such as carbon buckyballs and nanotubes) have been proposed. However, since Hydrogen usually amounts up to ~3.0-7.0 wt% at 77K which is far from the value set by US department of Energy (6 wt% at nearly ambient conditions), it makes carbon materials poor candidates for hydrogen storage.

## Clathrate Hydrates

$H_2$ caged in a clathrate hydrate was first reported in 2002, but requires very high pressures to be stable. In 2004, researchers from Delft University of Technology and Colorado School of Mines showed solid $H_2$-containing hydrates could be formed at ambient temperature and 10s of bar by adding small amounts of promoting substances such as THF. These clathrates have a theoretical maximum hydrogen densities of around 5 wt% and 40 kg/m³.

## Glass Capillary Arrays

A team of Russian, Israeli and German scientists have collaboratively developed an innovative technology based on glass capillary arrays for the safe infusion, storage and controlled release of hydrogen in mobile applications. The C.En technology has achieved the United States Department of Energy (DOE) 2010 targets for on-board hydrogen storage systems. DOE 2015 targets can be achieved using flexible glass capillaries and cryo-compressed method of hydrogen storage.

## Glass Microspheres

Hollow glass microspheres (HGM) can be utilized for controlled storage and release of hydrogen.

## Stationary Hydrogen Storage

Unlike mobile applications, hydrogen density is not a huge problem for stationary applications. As for mobile applications, stationary applications can use established technology:

- Compressed hydrogen ($CGH_2$) in a hydrogen tank

- Liquid hydrogen in a ($LH_2$) cryogenic hydrogen tank

- Slush hydrogen in a cryogenic hydrogen tank

## Underground Hydrogen Storage

Underground hydrogen storage is the practice of hydrogen storage in underground caverns, salt domes and depleted oil and gas fields. Large quantities of gaseous hydrogen have been stored in underground caverns by ICI for many years without any difficulties. The storage of large quantities of liquid hydrogen underground can function as grid energy storage. The round-trip efficiency is approximately 40% (vs. 75-80% for pumped-hydro (PHES)), and the cost is slightly higher than pumped hydro. The European project Hyunder indicated in 2013 that for the storage of wind and solar energy an additional 85 caverns are required as it can't be covered by PHES and CAES systems.

## Power to Gas

Power to gas is a technology which converts electrical power to a gas fuel. There are two methods: the first is to use the electricity for water splitting and inject the resulting hydrogen into the natural gas grid; the second, less efficient method is used to convert carbon dioxide and hydrogen to methane, using electrolysis and the Sabatier reaction. The excess power or off peak power generated by wind generators or solar arrays is then used for load balancing in the energy grid. Using the existing natural gas system for hydrogen Fuel cell maker Hydrogenics and natural gas distributor Enbridge have teamed up to develop such a power to gas system in Canada.

Pipeline storage of hydrogen where a natural gas network is used for the storage of hydrogen. Before switching to natural gas, the German gas networks were operated using towngas, which for the most part (60-65%) consisted of hydrogen. The storage capacity of the German natural gas network is more than 200,000 GW·h which is enough for several months of energy requirement. By comparison, the capacity of all German pumped storage power plants amounts to only about 40 GW·h. The transport of energy through a gas network is done with much less loss (<0.1%) than in a power network (8%). The use of the existing natural gas pipelines for hydrogen was studied by NaturalHy.

## Automotive Onboard Hydrogen Storage

| Storage Parameter | 2005 | 2010 | 2015 |
|---|---|---|---|
| Gravimetric Capacity (Specific energy) | 1.5 kWh/kg<br>0.045 kg $H_2$/kg | 2.0 kWh/kg<br>0.060 kg $H_2$/kg | 3.0 kWh/kg<br>0.090 kg $H_2$/kg |
| *System Weight:* | *111 Kg* | *83 Kg* | *55.6 Kg* |
| Volumetric Capacity (Energy density) | 1.2 kWh/L<br>0.036 kg $H_2$/L | 1.5 kWh/L<br>0.045 kg $H_2$/L | 2.7 kWh/L<br>0.081 kg $H_2$/L |
| *System Volume:* | *139 L* | *111 L* | *62 L* |
| Storage system cost | $6 /kWh | $4 /kWh | $2 /kWh |
| *System Cost:* | *$1000* | *$666* | *$333* |
| Refueling rate | .5 Kg $H_2$/min | 1.5 Kg $H_2$/min | 2.0 Kg $H_2$/min |
| *Refueling Time:* | *10 min* | *3.3 min* | *2.5 min* |

Targets were set by the FreedomCAR Partnership in January 2002 between the United States Council for Automotive Research (USCAR) and U.S. DOE (Targets assume a 5-kg $H_2$ storage system). The 2005 targets were not reached in 2005. The targets were revised in 2009 to reflect new data on system efficiencies obtained from fleets of test cars. The ultimate goal for volumetric storage is still above the theoretical density of liquid hydrogen.

It is important to note that these targets are for the hydrogen storage system, not the hydrogen storage material. System densities are often around half those of the working material, thus while a material may store 6 wt% $H_2$, a working system using that material may only achieve 3 wt% when the weight of tanks, temperature and pressure control equipment, etc., is considered.

In 2010, only two storage technologies were identified as having the potential to meet DOE targets: MOF-177 exceeds 2010 target for volumetric capacity, while cryo-compressed $H_2$ exceeds more restrictive 2015 targets for both gravimetric and volumetric capacity.

## Pumped-storage Hydroelectricity

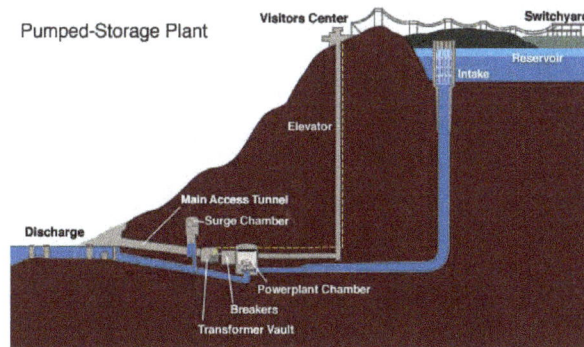

Diagram of the TVA pumped storage facility at Raccoon Mountain Pumped-Storage Plant

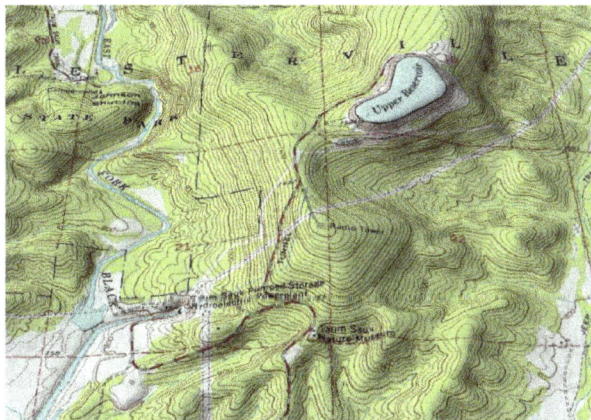

Shaded-relief topo map of the Taum Sauk pumped storage plant in Missouri

Pumped-storage hydroelectricity (PSH), or pumped hydroelectric energy storage (PHES), is a type of hydroelectric energy storage used by electric power systems for load balancing. The method stores energy in the form of gravitational potential energy of water, pumped from a lower elevation reservoir to a higher elevation. Low-cost surplus off-peak electric power is typically used to run the pumps. During periods of high electrical demand, the stored water is released through turbines to produce electric power. Although the losses of the pumping process makes the plant a net consumer of energy overall, the system increases revenue by selling more electricity during periods of *peak demand*, when electricity prices are highest.

Pumped-storage hydroelectricity allows energy from intermittent sources (such as solar, wind) and other renewables, or excess electricity from continuous base-load sources (such as coal or nuclear) to be saved for periods of higher demand. The reservoirs used with pumped storage are quite small when compared to conventional hydroelectric dams of similar power capacity, and generating periods are often less than half a day.

Pumped storage is the largest-capacity form of grid energy storage available, and, as of 2017, the DOE Global Energy Storage Database reports that PSH accounts for over 96% of all active tracked storage installations worldwide, with a total installed nameplate capacity of over 168 GW. The *round-trip* energy efficiency of PSH varies between 70%–80%, with some sources claiming up to 87%. The main disadvantage of PHS is the specialist nature of the site required, needing both geographical height and water availability. Suitable sites are therefore likely to be in hilly or mountainous regions, and potentially in areas of outstanding natural beauty, and therefore there are also social and ecological issues to overcome.

## Overview

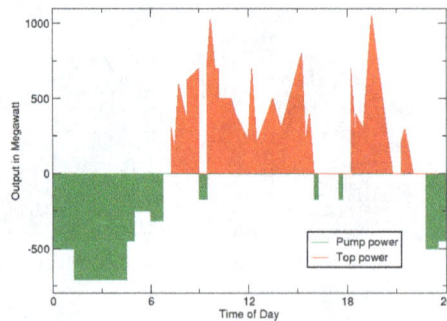

Power distribution, over a day, of a pumped-storage hydroelectricity facility.
Green represents power consumed in pumping; red is power generated.

At times of low electrical demand, excess generation capacity is used to pump water into the upper reservoir. When there is higher demand, water is released back into the lower reservoir through a turbine, generating electricity. Reversible turbine/generator assemblies act as a combined pump and turbine generator unit (usually a Francis turbine design). In open-loop systems, pure pumped-storage plants store water in an upper reservoir with no natural inflows, while pump-back plants utilize a combination of pumped storage and conventional hydroelectric plants with an upper reservoir that is replenished in part by natural inflows from a stream or river. Plants that do not use pumped-storage are referred to as conventional hydroelectric plants; conventional hydroelectric plants that have significant storage capacity may be able to play a similar role in the electrical grid as pumped storage by deferring output until needed.

Taking into account evaporation losses from the exposed water surface and conversion losses, energy recovery of 70-80% or more can be regained. This technique is currently the most cost-effective means of storing large amounts of electrical energy, but capital costs and the presence of appropriate geography are critical decision factors in selecting pumped-storage plant sites.

The relatively low energy density of pumped storage systems requires either large flows and/or large differences in height between reservoirs. The only way to store a significant amount of energy is by having a large body of water located relatively near, but as high above as possible, a second body of water. In some places this occurs naturally, in others one or both bodies of water were man-made. Projects in which both reservoirs are artificial and in which no natural inflows are involved with either reservoir are referred to as "closed loop" systems.

These systems may be economical because they flatten out load variations on the power grid, permitting thermal power stations such as coal-fired plants and nuclear power plants that provide base-load electricity to continue operating at peak efficiency, while reducing the need for "peaking" power plants that use the same fuels as many base-load thermal plants, gas and oil, but have been designed for flexibility rather than maximal efficiency. Hence pumped storage systems are crucial when coordinating large groups of heterogeneous generators. However, capital costs for pumped-storage plants are relatively high.

The upper reservoir (Llyn Stwlan) and dam of the Ffestiniog Pumped Storage Scheme in north Wales. The lower power station has four water turbines which generate 360 MW of electricity within 60 seconds of the need arising.

Along with energy management, pumped storage systems help control electrical network frequency and provide reserve generation. Thermal plants are much less able to respond to sudden changes in electrical demand, potentially causing frequency and voltage instability. Pumped storage plants, like other hydroelectric plants, can respond to load changes within seconds.

The most important use for pumped storage has traditionally been to balance baseload powerplants, but may also be used to abate the fluctuating output of intermittent energy sources. Pumped storage provides a load at times of high electricity output and low electricity demand, enabling additional system peak capacity. In certain jurisdictions, electricity prices may be close to zero or occasionally negative on occasions that there is more electrical generation available than there is load available to absorb it; although at present this is rarely due to wind or solar power alone, increased wind and solar generation will increase the likelihood of such occurrences. It is particularly likely that pumped storage will become especially important as a balance for very large scale photovoltaic generation. Increased long distance transmission capacity combined with significant amounts of energy storage will be a crucial part of regulating any large-scale deployment of intermittent renewable power sources. The high non-firm renewable electricity penetration in some regions supplies 40% of annual output, but 60% may be reached before additional storage is necessary.

There are small-scale installations of such technology, namely in buildings, although these are economically unfeasible given the economies of scale present. Also, a large volume of water is required for a meaningful storage capacity which is a difficult fit for an urban setting. Nevertheless, some authors defend its technological simplicity and secure provision of water as important externalities.

## History

The first use of pumped storage was in the 1890s in Italy and Switzerland. In the 1930s reversible hydroelectric turbines became available. These turbines could operate as both turbine-generators and in reverse as electric motor driven pumps. The latest in large-scale engineering technology are variable speed machines for greater efficiency. These machines operate in synchronization with the network frequency when generating, but operate asynchronously (independent of the network frequency) when pumping.

The first use of pumped-storage in the United States was in 1930 by the Connecticut Electric and Power Company, using a large reservoir located near New Milford, Connecticut, pumping water from the Housatonic River to the storage reservoir 230 feet above.

## Worldwide use

In 2009, world pumped storage generating capacity was 104 GW, while other sources claim 127 GW, which comprises the vast majority of all types of utility grade electric storage. The EU had 38.3 GW net capacity (36.8% of world capacity) out of a total of 140 GW of hydropower and representing 5% of total net electrical capacity in the EU. Japan had 25.5 GW net capacity (24.5% of world capacity).

In 2010 the United States had 21.5 GW of pumped storage generating capacity (20.6% of world capacity). PHS generated (net) -5.501 GWh of energy in 2010 in the US because more energy is consumed in pumping than is generated. Nameplate pumped storage capacity had grown to 21.6 GW by 2014, with pumped storage comprising 97% of grid-scale energy storage in the US. As of late 2014, there were 51 active project proposals with a total of 39 GW of new nameplate capacity across all stages of the FERC licensing process for new pumped storage hydroelectric plants in the US, but no new plants were currently under construction in the US at the time.

The five largest operational pumped-storage plants are listed below:

| Station | Country | Location | Capacity (MW) |
|---|---|---|---|
| Bath County Pumped Storage Station | United States | 38°12′32″N 79°48′00″W / 38.20889°N 79.80000°W / 38.20889; -79.80000 (Bath County Pumped-storage Station) | 3,003 |
| Guangdong Pumped Storage Power Station | China | 23°45′52″N 113°57′12″E / 23.76444°N 113.95333°E / 23.76444; 113.95333 (Guangzhou Pumped Storage Power Station) | 2,400 |
| Huizhou Pumped Storage Power Station | China | 23°16′07″N 114°18′50″E / 23.26861°N 114.31389°E / 23.26861; 114.31389 (Huizhou Pumped Storage Power Station) | 2,400 |
| Okutataragi Pumped Storage Power Station | Japan | 35°14′13″N 134°49′55″E / 35.23694°N 134.83194°E / 35.23694; 134.83194 (Okutataragi Hydroelectric Power Station) | 1,932 |

| Ludington Pumped Storage Power Plant | United States | 43°53′37″N 86°26′43″W / 43.89361°N 86.44528°W / 43.89361; -86.44528 (Ludington Pumped Storage Power Plant) | 1,872 |

Note: this table shows the power-generating capacity in megawatts as is usual for power stations. However, the overall energy-storage capacity in megawatt-hours (MWh) is a different intrinsic property and can not be derived from the above given figures.

## Pump-back Hydroelectric Dams

Conventional hydroelectric dams may also make use of pumped storage in a hybrid system that both generates power from water naturally flowing into the reservoir as well as storing water pumped back to the reservoir from below the dam. The Grand Coulee Dam in the US was expanded with a pump-back system in 1973. Existing dams may be repowered with reversing turbines thereby extending the length of time the plant can operate at capacity. Optionally a pump back powerhouse such as the Russell Dam (1992) may be added to a dam for increased generating capacity. Making use of an existing dams upper reservoir and transmission system can expedite projects and reduce costs.

## Potential Technologies

### Seawater

Pumped storage plants can operate with seawater, although there are additional challenges compared to using fresh water. In 1999, the 30 MW Yanbaru project in Okinawa was the first demonstration of seawater pumped storage. The 300 MW seawater-based Lanai Pumped Storage Project has recently been proposed on Lanai, Hawaii, and seawater-based projects have been proposed in Ireland. A pair of proposed projects in the Atacama Desert in northern Chile would use 600 MW of photovoltaic solar (Skies of Tarapacá) together with 300 MW of pumped storage (Mirror of Tarapacá) raising seawater 600 metres (2,000 ft) up a coastal cliff.

### Underground Reservoirs

The use of underground reservoirs has been investigated. Recent examples include the proposed Summit project in Norton, Ohio, the proposed Maysville project in Kentucky (underground limestone mine), and the Mount Hope project in New Jersey, which was to have used a former iron mine as the lower reservoir. Several new underground pumped storage projects have been proposed. Cost-per-kilowatt estimates for these projects can be lower than for surface projects if they use existing underground mine space. There are limited opportunities involving suitable underground space, but the number of underground pumped storage opportunities may increase if abandoned coal mines prove suitable.

### Decentralised Systems

Small pumped-storage hydropower plants can be built on streams and within infrastructures, such as drinking water networks and artificial snow making infrastructures. Such plants provide distributed energy storage and distributed flexible electricity production and can contribute to the decentralized integration of intermittent renewable energy technologies, such as wind power and solar power. Reservoirs that can be used for small pumped-storage hydropower plants could include natural or artificial lakes, reservoirs within other structures such as irrigation, or unused portions of mines or underground military installations. In Switzerland one study suggested that the total installed capacity of small pumped-storage hydropower plants in 2011 could be increased by 3 to 9 times by providing adequate policy instruments.

## Underwater Reservoirs

In March 2017 the research project StEnSea (Storing Energy at Sea) announced their successful completion of a four-week test of a pumped storage underwater reservoir. In this configuration a hollow sphere submerged in deep water acts as the lower reservoir while the upper reservoir is the enclosing body of water. When a reversible turbine integrated into the sphere uses surplus electricity to pump water out of the sphere the force of the pump must act on the entire column of water above the sphere, so the deeper the sphere is located, the more potential energy it can store and convert back to electricity by letting water back in via the turbine.

As such the energy storage capacity of the submerged reservoir is not governed by the gravitational energy in the traditional sense, but rather by the vertical pressure variation.

While StEnSea's test took place at a depth of 100 m in the fresh water Lake Constance, the technology is foreseen to be used in salt water at greater depths. Since the submerged reservoir needs only a connecting electrical cable, the depth at which it can be employed is limited only by the depth at which the turbine can function, currently limited to 700 m. The challenge of designing salt water pumped storage in this underwater configuration brings a range of advantages:

- No land area is required,

- No mechanical structure other than the electrical cable needs to span the distance of the potential energy difference,

- In the presence of sufficient seabed area multiple reservoirs can scale the storage capacity without limits,

- Should a reservoir collapse, the consequences would be limited apart from the loss of the reservoir itself,

- Evaporation from the upper reservoir has no effect on the energy conversion efficiency,

- Transmission of electricity between the reservoir and the grid can be established from a nearby offshore wind farm limiting transmission loss and obviating the need for onshore cabling permits.

A current commercial design featuring a sphere with an inner diameter of 30 m submerged to 700 m would correspond to a 20 MWh capacity which with a 5 MW turbine would lead to a 4-hour discharge time. An energy park with multiple such reservoirs would bring the storage cost to around a few eurocents per kWh with construction and equipment costs in the range €1,200-€1,400 per kW. To avoid excessive transmission cost and loss, the reservoirs should be placed off deep water coasts of densely populated areas, such as Norway, Spain, USA and Japan. With this limitation the concept would allow for worldwide electricity storage of close to 900 GWh.

For comparison, a traditional, gravity-based pumped storage capable of storing 20 MWh in a water reservoir the size of a 30 m sphere would need a hydraulic head of 519 m with the elevation spanned by a pressurized water pipe requiring typically a hill or mountain for support.

## Underground Hydrogen Storage

Underground hydrogen storage is the practice of hydrogen storage in underground caverns, salt domes and depleted oil/gas fields. Large quantities of gaseous hydrogen have been stored in underground caverns by ICI for many years without any difficulties. The storage of large quantities of hydrogen underground in solution-mined salt domes, aquifers or excavated rock caverns or mines can function as grid energy storage which is essential for the hydrogen economy. By using a turboexpander the electricity needs for compressed storage on 200 bar amounts to 2.1% of the energy content.

## Chevron Phillips Clemens Terminal

The Chevron Phillips Clemens Terminal in Texas has stored hydrogen since the 1980s in a solution-mined salt cavern. The cavern roof is about 2,800 feet (850 m) underground. The cavern is a cylinder with a diameter of 160 feet (49 m), a height of 1,000 feet (300 m) and a usable hydrogen capacity of 1,066 million cubic feet ($30.2 \times 10^6$ m³), or 2,520 metric tons (2,480 long tons; 2,780 short tons).

## Development

- Sandia National Laboratories released in 2011 a life cycle cost analysis framework for geologic storage of hydrogen.

- The European project Hyunder indicated in 2013 that for the storage of wind and solar energy an additional 85 caverns are required as it can't be covered by PHES and CAES systems.

- ETI released in 2015 a report *The role of hydrogen storage in a clean responsive power system* noting that the UK has sufficient salt bed resources to provide tens of GWe.

# Thermal Energy Storage

Thermal energy storage (TES) is achieved with greatly differing technologies that collectively accommodate a wide range of needs. It allows excess thermal energy to be collected for later use, hours, days or many months later, at individual building, multiuser building, district, town, or even regional scale depending on the specific technology. As examples: energy demand can be balanced between daytime and nighttime; summer heat from solar collectors can be stored interseasonally for use in winter; and cold obtained from winter air can be provided for summer air conditioning. Storage media include: water or ice-slush tanks ranging from small to massive, masses of native earth or bedrock accessed with heat exchangers in clusters of small-diameter boreholes (sometimes quite deep); deep aquifers contained between impermeable strata; shallow, lined pits filled with gravel and water and top-insulated; and eutectic, phase-change materials.

District heating accumulation tower from Theiss near Krems
an der Donau in Lower Austria with a thermal capacity of 2 GWh.

Other sources of thermal energy for storage include heat or cold produced with heat pumps from off-peak, lower cost electric power, a practice called peak shaving; heat from combined heat and power (CHP) power plants; heat produced by renewable electrical energy that exceeds grid demand and waste heat from industrial processes. Heat storage, both seasonal and short term, is considered an important means for cheaply balancing high shares of variable renewable electricity production and integration of electricity and heating sectors in energy systems almost or completely fed by renewable energy.

## Solar Energy Storage

Most practical active solar heating systems provide storage from a few hours to a day's worth of energy collected. However, there are a growing number of facilities that use seasonal thermal energy storage (STES), enabling solar energy to be stored in summer for space heating use during winter. The Drake Landing Solar Community in Alberta, Canada, has now achieved a year-round 97% solar heating fraction, a world record made possible only by incorporating STES.

The use of both latent heat and sensible heat are also possible with high temperature solar thermal input. Various eutectic mixtures of metals, such as Aluminium and Silicon (AlSi12) offer a high melting point suited to efficient steam generation, while high alumina cement-based materials offer good thermal storage capabilities.

Molten salt is a means of storing solar energy at a high temperature.

## Molten Salt Technology

Molten salts can be employed as a thermal energy storage method to retain thermal energy. This is a current commercially used technology to store the heat collected by concentrated solar power (e.g., from a solar tower or solar trough). The heat can later be converted into superheat steam to power conventional steam turbines and generate electricity in bad weather or at night. It was demonstrated in the Solar Two project from 1995-1999. Estimates in 2006 predicted an annual efficiency of 99%, a reference to the energy retained by storing heat before turning it into electricity, versus converting heat directly into electricity. Various eutectic mixtures of different salts are used (e.g., sodium nitrate, potassium nitrate and calcium nitrate). Experience with such systems exists in non-solar applications in the chemical and metals industries as a heat-transport fluid.

The salt melts at 131 °C (268 °F). It is kept liquid at 288 °C (550 °F) in an insulated "cold" storage tank. The liquid salt is pumped through panels in a solar collector where the focused sun heats it to 566 °C (1,051 °F). It is then sent to a hot storage tank. With proper insulation of the tank the thermal energy can be usefully stored for up to a week.

When electricity is needed, the hot salt is pumped to a conventional steam-generator to produce superheated steam for a conventional turbine/generator set as used in any coal, oil, or nuclear power plant. A 100-megawatt turbine would need a tank of about 9.1 metres (30 ft) tall and 24 metres (79 ft) in diameter to drive it for four hours by this design.

Several parabolic trough power plants in Spain and solar power tower developer SolarReserve use this thermal energy storage concept. The Solana Generating Station in the U.S. can store 6 hours worth of generating capacity in molten salt. During the summer of 2013 the Gemasolar Thermo-solar solar power tower/molten salt plant in Spain achieved a first by continuously producing electricity 24 hours per day for 36 days.

## Heat Storage in Tanks or Rock Caverns

A steam accumulator consists of an insulated steel pressure tank containing hot water and steam under pressure. As a heat storage device, it is used to mediate heat production by a variable or steady source from a variable demand for heat. Steam accumulators may take on a significance for energy storage in solar thermal energy projects.

Large stores are widely used in Scandinavia to store heat for several days, to decouple heat and power production and to help meet peak demands. Interseasonal storage in caverns has been investigated and appears to be economical.

## Heat Storage in Hot Rocks, Concrete, Pebbles etc

Water has one of the highest thermal capacities Heat capacity - 4.2 $J/(cm^3{\cdot}K)$ whereas concrete has about one third of that. On the other hand, concrete can be heated to much higher temperatures – 1200 °C by e.g. electrical heating and therefore has a much higher overall volumetric capacity. Thus in the example below, an insulated cube of about 2.8 m would appear to provide sufficient storage for a single house to meet 50% of heating demand. This could, in principle, be used to store surplus wind or PV heat due to the ability of electrical heating to reach high temperatures. At the neighborhood level, the Wiggenhausen-Süd solar development at Friedrichshafen has received international attention. This features a 12,000 $m^3$ (420,000 cu ft) reinforced concrete thermal store linked to 4,300 $m^2$ (46,000 sq ft) of solar collectors, which will supply the 570 houses with around 50% of their heating and hot water. Siemens builds a 36 MWh thermal storage near Hamburg with 600 °C basalt and 1.5 MW electric output. A similar system is scheduled for Sorø, Denmark, with 41-58% of the stored 18 MWh heat returned for the town's district heating, and 30-41% returned as electricity.

## Miscibility Gap Alloy Technology (MGA)

Miscibility gap alloys rely on the phase change of a metallic material to store thermal energy.

Rather than pumping the liquid metal between tanks as in a molten salt system, the metal is encapsulated in another metallic material that it cannot alloy with (immiscible). Depending on the two materials selected (the phase changing material and the encapsulating material) storage densities can be between 0.2 and 2 MJ/L.

A working fluid, typically water or steam, is used to transfer the heat into and out of the MGA. Thermal conductivity of MGAs is often higher (up to 400 W/m K) than competing technologies which means quicker "charge" and "discharge" of the thermal storage is possible. The technology has not yet been implemented on a large scale.

## Electric Thermal Storage Heaters

Storage heaters are commonplace in European homes with time-of-use metering (traditionally using cheaper electricity at night time). They consist of high-density ceramic bricks or feolite blocks heated to a high temperature with electricity, and may or may not have good insulation and controls to release heat over a number of hours.

## Ice-based Technology

Several applications are being developed where ice is produced during off-peak periods and used for cooling at later time. For example, air conditioning can be provided more economically by using low-cost electricity at night to freeze water into ice, then using the cooling capacity of ice in the afternoon to reduce the electricity needed to handle air conditioning demands. Thermal energy storage using ice makes use of the large heat of fusion of water. Historically, ice was transported from mountains to cities for use as a coolant. One metric ton of water (= one cubic meter) can store 334 million joules (MJ) or 317,000 BTUs (93kWh). A relatively small storage facility can hold enough ice to cool a large building for a day or a week.

In addition to using ice in direct cooling applications, it is also being used in heat pump based heating systems. In these applications the phase change energy provides a very significant layer of thermal capacity that is near the bottom range of temperature that water source heat pumps can operate in. This allows the system to ride out the heaviest heating load conditions and extends the timeframe by which the source energy elements can contribute heat back into the system.

## Cryogenic Energy Storage

This uses liquification of air or nitrogen as an energy store.

A pilot cryogenic energy system that uses liquid air as the energy store, and low-grade waste heat to drive the thermal re-expansion of the air, has been operating at a power station in Slough, UK since 2010.

## Hot Silicon Technology

Solid or molten silicon offers much higher storage temperatures than salts with consequent greater capacity and efficiency. It is being researched as a possible more energy efficient storage technology. Silicon is able to store more than 1MWh of energy per cubic metre at 1400 °C.

## Pumped-heat Electricity Storage

In pumped-heat electricity storage (PHES), a reversible heat-pump system is used to store energy as a temperature difference between two heat stores.

## Isentropic

One system which was being developed by the now bankrupt UK company Isentropic operates as follows. It comprises two insulated containers filled with crushed rock or gravel; a hot vessel storing thermal energy at high temperature and high pressure, and a cold vessel storing thermal energy at low temperature and low pressure. The vessels are connected at top and bottom by pipes and the whole system is filled with the inert gas argon.

During the charging cycle the system uses off-peak electricity to work as a heat pump. Argon at ambient temperature and pressure from the top of the cold store is compressed adiabatically to a pressure of 12 bar, heating it to around 500 °C (900 °F). The compressed gas is transferred to the top of the hot vessel where it percolates down through the gravel, transferring its heat to the rock and cooling to ambient temperature. The cooled, but still pressurized, gas emerging at the bottom of the vessel is then expanded (again adiabatically) back down to 1 bar, which lowers its temperature to -150 °C. The cold gas is then passed up through the cold vessel where it cools the rock while being warmed back to its initial condition.

The energy is recovered as electricity by reversing the cycle. The hot gas from the hot vessel is expanded to drive a generator and then supplied to the cold store. The cooled gas retrieved from the bottom of the cold store is compressed which heats the gas to ambient temperature. The gas is then transferred to the bottom of the hot vessel to be reheated.

The compression and expansion processes are provided by a specially designed reciprocating machine using sliding valves. Surplus heat generated by inefficiencies in the process is shed to the environment through heat exchangers during the discharging cycle.

The developer claims that a round trip efficiency of 72-80% is achievable. This compares to >80% achievable with pumped hydro energy storage.

Another proposed system uses turbomachinery and is capable of operating at much higher power levels.

## Endothermic/Exothermic Chemical Reactions

### Salt Hydrate Technology

One example of an experimental storage system based on chemical reaction energy is the salt hydrate technology. The system uses the reaction energy created when salts are hydrated or dehydrated. It works by storing heat in a container containing 50% sodium hydroxide (NaOH) solution. Heat (e.g. from using a solar collector) is stored by evaporating the water in an endothermic reaction. When water is added again, heat is released in an exothermic reaction at 50 °C (120 °F). Current systems operate at 60% efficiency. The system is especially advantageous for seasonal thermal energy storage, because the dried salt can be stored at room temperature for prolonged times, without energy loss. The containers with the dehydrated salt can even be transported to a different location. The system has a higher energy density than heat stored in water and the capacity of the system can be designed to store energy from a few months to years.

In 2013 the Dutch technology developer TNO presented the results of the MERITS project to store heat in a salt container. The heat, which can be derived from a solar collector on a rooftop, expels the water contained in the salt. When the water is added again, the heat is released, with almost no energy losses. A container with a few cubic meters of salt could store enough of this thermo-chemical energy to heat a house throughout the winter. In a temperate climate like that of the Netherlands, an average low-energy household requires about 6.7 GJ/winter. To store this energy in water (at a temperature difference of 70 °C), 23 m$^3$ insulated water storage would be needed, exceeding the storage abilities of most households. Using salt hydrate technology with a storage density of about 1 GJ/m$^3$, 48 m$^3$ could be sufficient.

As of 2016, researchers in several countries are conducting experiments to determine the best type of salt, or salt mixture. Low pressure within the container seems favourable for the energy transport. Especially promising are organic salts, so called ionic liquids. Compared to lithium halide based sorbents they are less problematic in terms of limited global ressources, and compared to most other halides and sodium hydroxide (NaOH) they are less corrosive and not negatively affected by $CO_2$ contaminations.

## Molecular Bonds

Storing energy in molecular bonds is being investigated. Energy densities equivalent to lithium-ion batteries have been achieved.

## Aquifer Thermal Energy Storage

Aquifer thermal energy storage (ATES) is the storage and recovery of thermal energy in the subsurface. ATES is applied to provide heating and cooling to buildings. Storage and recovery of thermal energy is achieved by extraction and injection of groundwater from aquifers using groundwater wells. Systems commonly operate in a seasonal mode. The groundwater that is extracted in summer, is used for cooling by transferring heat from the building to the groundwater by means of a heat exchanger. Subsequently, the heated groundwater is injected back into the aquifer, which creates a storage of heated groundwater. In wintertime, the flow direction is reversed such that the heated groundwater is extracted and can be used for heating (often in combination with a heat pump). Therefore, operating an ATES system uses the subsurface as a temporal storage to buffer seasonal variations in heating and cooling demand. When replacing traditional fossil fuel dependent heating and cooling systems, ATES can serve as a cost-effective technology to reduce the primary energy consumption of a building and the associated $CO_2$ emissions.

In 2009 United Nations Climate Change Conference in Copenhagen, Denmark, many countries and regions have made targets for global climate protection. The European Union also set a target to reduce greenhouse gas emissions, increase use of sustainable energy and improve energy efficiency. For this target, ATES can actually contribute significantly, as about 40% of global energy consumption is done by buildings, and is mainly for heating and cooling. Therefore, the development of ATES has been paid a lot of attention and the number of ATES has increased dramatically, especially in Europe. For example, in the Netherlands, it was estimated that about 20,000 ATES systems could be achieved by 2020. This can yield a $CO_2$ emission reduction of about 11%, for the target of the Netherlands. Besides the Netherlands, Belgium, Germany, Turkey, and Sweden are also increasing the application of ATES. ATES can be applied worldwide, as long as the climatic conditions and geohydrological conditions are right. As ATES systems cumulate in urban areas optimisation of subsurface space requires attention in areas with suitable conditions.

## System Types

In its basic form, an ATES system consists of two wells (called a doublet). One well is used for heat storage, and the other for cold storage. During winter, (warm) groundwater is extracted from the heat storage well and injected in the cold storage well. During summer, the flow direction is reversed such that (cold) groundwater is extracted from the cold storage well and injected in the heat storage well. Because each well serves both as an extraction and injection well, these systems are called bi-directional. There are also mono-directional systems. These systems do not switch pumping direction, such that groundwater is always extracted at the natural aquifer temperature. Although thermal energy is stored in the subsurface, there is usually no intention to retrieve the stored energy.

Thermal energy storage can also be achieved by circulating a fluid through a buried heat exchanger, that usually consists of a horizontal or vertical pipeline. As these systems do not extract or inject groundwater, they are called closed systems and are known as borehole thermal energy storage or ground source heat pumps. Another thermal application that uses the subsurface to provide thermal energy is geothermal energy production, which commonly uses the deeper subsurface where temperature is higher.

## History

The first reported deliberate storage of thermal energy in aquifers was in China around 1960. There, large amounts of groundwater were extracted to supply cooling to an industrial facility. This led to substantial land subsidence. To inhibit the subsidence, cold surface water was injected back into the aquifer. Subsequently, it was observed that the stored water remained cold after injection and could be used for industrial cooling. Storage of thermal energy in aquifers was further suggested in the 1970s which led to field experiments and feasibility studies in France, Switzerland, US and Japan. There are no official statistics on the number and size of ATES systems worldwide. However, the Netherlands and Sweden are considered to dominate the market in terms of implementation. In Sweden there were approximately 104 ATES systems in 2012 with a total capacity of 110 MW. The number of ATES systems in the Netherlands in the same year was 2740, with a total estimated capacity of 1103 MW.

## Typical Dimensions

Flow rates for typical applications in the utility sector are between 20 and 150 m³/hour for each well. The total volume of groundwater that is stored and recovered in a year generally varies between 10 000 m³ and 150 000 m³ per well. Depth at which ATES is applied varies commonly between 20 and 200 meters below surface. Temperature at these depths is generally close to the annual mean surface temperature. In moderate climates this is around 10 °C. In those regions cold storage is commonly applied between 5 and 10 °C and heat storage in the range 10 to 20 °C. Although less frequent, there are also some projects reported in which heat was stored above 80 °C.

## Hydrogeological Constrains

Energy savings that can be achieved with ATES are strongly depending on the geology of a site. Mainly, ATES requires the presence of a suitable aquifer that is able to accept and yield water. Therefore, thick (>10 m) sandy aquifers are selected. Natural groundwater flow may transport (part of) the stored energy outside of the capture zone of a well during the storage phase. To reduce advective heat loss, aquifers with a low hydraulic gradient are preferred. In addition, gradients in geochemical composition should be avoided, as mixing of water with different geochemistry can increase clogging, which will reduce the performance of a well and lead to increased maintenance costs.

## Legal Status

The legal status of shallow geothermal installations (<400 m) is diverse among countries. Regulations for installations of wells concern the use of hazardous materials and proper backfilling of the drilling hole to avoid hydraulic short circuiting between aquifers. Other legislation concerns protection of groundwater areas for drinking water supply. Some countries adopt limits for minimum and maximum storage temperatures. For example, Austria (5–20 °C), Denmark (2–25 °C) and Netherlands (5–25 °C). While other countries adopt a maximum change in groundwater temperature, for example Switzerland (3 °C) and France (11 °C).

## Interference with Chlorinated Ethenes (CVOCs)

ATES is currently not allowed to be applied in contaminated aquifers, due to the possible spreading of contaminants in the groundwater of the subsurface, especially in urban areas. This will lead to deterioration of the quality of groundwater, which is also an important source for drinking water. Despite the regulations made to prevent the interference between ATES and groundwater contaminants, the possibility of their encounter is however rising, because of the rapid increase of the number of ATES and slow progress of remediation groundwater contaminations in urban area. Among the common groundwater contaminants, chlorinated ethenes have a most chance to interfere with the ATES system, as they are often found at the similar depth as ATES. When chlorinated ethenes present as Dense non-aqueous phase liquid (DNAPLs), the possible dissolution of DNAPLs by ATES will cause more severe impact on the groundwater quality.

## Possible Application in Contaminated Area

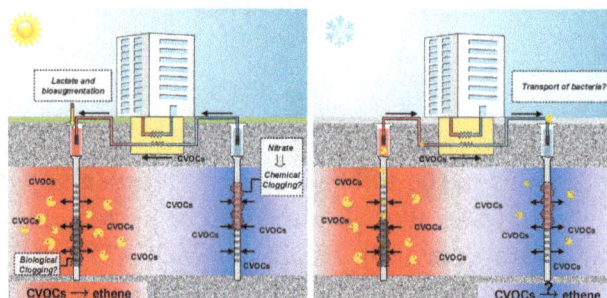

Illustration of relevant processes in the ATES-ENA system

The possible interference between ATES and chlorinated ethenes has been also seen as an opportunity of integration of sustainable energy technology and sustainable groundwater management. The combination of ATES and enhanced bioremediation first introduced in the "More with Sub-Surface Energy" (Meer met Bodemenergie, MMB) project in the Netherlands in 2009. Several scientific and practical rationales are the basics for seeing such combination as a promising possibility. Increased temperature around the warm well can enhance reductive dechlorination of chlorinated ethenes. Although low temperature in cold well can hamper the biodegradation, the seasonal operation of ATES can transfer contaminant from cold well to hot well for faster biodegradation. Such seasonal groundwater transportation can also homogenize the environmental condition. ATES can be used as a biostimulation too, for example to inject electron donor or microorganism needed for reductive dechlorination. Finally, the life time of ATES (30 years) fits the long duration of in situ bioremediation.

## Societal Impacts

The combination concept of ATES and enhanced natural attenuation (ATES-ENA) can possibly be used in the Netherlands and China, especially in urbanized areas. These areas in both countries are confronted with organic groundwater contaminations. Currently, the combination concept may be better applicable for the Netherlands with more mature technology and application of ATES. And the overlapping between ATES and groundwater contamination also promotes the need of this combined technology. However, for China where ATES is much less developed compared to the Netherlands, the important advantages are that many more demonstration pilot projects can be set-up prior to real applications, and flexible systems can be developed because of the less intense pressure on subsurface use by ATES compared to the Netherlands. For sustainable urban development, the combined ATES-ENA technology can provide contributions to the solution of both energy and environmental problems.

## Seasonal Thermal Energy Storage

Seasonal thermal energy storage (or STES) is the storage of heat or cold for periods of up to several months. The thermal energy can be collected whenever it is available and be used whenever needed, such as in the opposing season. For example, heat from solar collectors or waste heat from air conditioning equipment can be gathered in hot months for space heating use when needed, including during winter months. Waste heat from industrial process can similarly be stored and be used much later. Or the natural cold of winter air can be stored for summertime air conditioning. STES stores can serve district heating systems, as well as single buildings or complexes. Among seasonal storages used for heating, the design peak annual temperatures generally are in the range of 27 to 80 °C (81 to 180 °F), and the temperature difference occurring in the storage over the course of a year can be several tens of degrees. Some systems use a heat pump to help charge and discharge the storage during part or all of the cycle. For cooling applications, often only circulation pumps are used. A less common term for STES technologies is interseasonal thermal energy storage.

Examples for district heating include Drake Landing Solar Community where ground storage provides 97% of yearly consumption without heat pumps, and Danish pond storage with boosting.

## STES Technologies

There are several types of STES technology, covering a range of applications from single small buildings to community district heating networks. Generally, efficiency increases and the specific construction cost decreases with size.

## Underground Thermal Energy Storage

- UTES (underground thermal energy storage), in which the storage medium may be geological strata ranging from earth or sand to solid bedrock, or aquifers. UTES technologies include:

  o ATES (aquifer thermal energy storage). An ATES store is composed of a doublet, totaling two or more wells into a deep aquifer that is contained between impermeable geological layers above and below. One half of the doublet is for water extraction and the other half for reinjection, so the aquifer is kept in hydrological balance, with no net extraction. The heat (or cold) storage medium is the water and the substrate it occupies. Germany's Reichstag building has been both heated and cooled since 1999 with ATES stores, in two aquifers at different depths. In the Netherlands there are now well over 1,000 ATES systems, which are now a standard construction option. A significant system has been operating at Richard Stockton College (New Jersey) for several years. ATES has a lower installation cost than BTES because usually fewer holes are drilled, but ATES has a higher operating cost. Also, ATES requires particular underground conditions to be feasible, including the presence of an aquifer.

  o BTES (borehole thermal energy storage). BTES stores can be constructed wherever boreholes can be drilled, and are composed of one to hundreds of vertical boreholes, typically 155 mm (6.102 in) in diameter. Systems of all sizes have been built, including many quite large. The strata can be anything from sand to crystalline hardrock, and depending on engineering factors the depth can be from 50 to 300 metres (164 to 984 ft). Spacings have ranged from 3 to 8 metres (9.8 to 26.2 ft). Thermal models can be used to predict seasonal temperature variation in the ground, including the establishment of a stable temperature regime which is achieved by matching the inputs and outputs of heat over one or more annual cycles. Warm-temperature seasonal heat stores can be created using borehole fields to store surplus heat captured in summer to actively raise the temperature of large thermal banks of soil so that heat can be extracted more easily (and more cheaply) in winter. Interseasonal Heat Transfer uses water circulating in pipes embedded in asphalt solar collectors to transfer heat to Thermal Banks created in borehole fields. A ground source heat pump is used in winter to extract the warmth from the Thermal Bank to provide space heating via underfloor heating. A high Coefficient of Performance is obtained because the heat pump starts with a warm temperature of 25 °C (77 °F) from the thermal store, instead of a cold temperature of 10 °C (50 °F) from the ground. A BTES operating at Richard Stockton College since 1995 at a peak of about 29 °C (84.2 °F) consists of 400 boreholes 130 metres (427 ft) deep under a 3.5-acre (1.4 ha) parking lot. It has a heat loss of 2% over six months. The upper temperature limit for a BTES store is 85 °C (185 °F) due to characteristics of the PEX pipe used for BHEs, but most do not approach that limit. Boreholes can be either grout- or water-filled depending on geological conditions, and usually have a life expectancy in excess of 100 years. Both a BTES and its associated district heating system can be expanded incrementally after operation begins, as at Neckarsulm, Germany. BTES stores generally do not impair use of the land, and can exist under buildings, agricultural fields and parking lots. An example of one of the several kinds of STES illustrates well the capability of interseasonal heat storage. In Alberta, Canada, the homes of the Drake Landing Solar Community (in operation since 2007), get 97% of their year-round heat from a district heat system that is supplied by solar heat from solar-thermal panels on garage roofs. This feat – a world record – is enabled by interseasonal heat storage in a large mass of native rock that is under a central park. The thermal exchange occurs via a cluster of 144 boreholes, drilled 37 metres (121 ft) into the earth. Each borehole is 155 mm (6.1 in) in diameter and contains a simple heat exchanger made of small diameter plastic pipe, through which water is circulated. No heat pumps are involved.

o   CTES (cavern or mine thermal energy storage). STES stores are possible in flooded mines, purpose-built chambers, or abandoned underground oil stores (e.g. those mined into crystalline hardrock in Norway), if they are close enough to a heat (or cold) source and market.

o   Energy Pilings. During construction of large buildings, BHE heat exchangers much like those used for BTES stores have been spiraled inside the cages of reinforcement bars for pilings, with concrete then poured in place. The pilings and surrounding strata then become the storage medium.

o   GIITS (geo interseasonal insulated thermal storage). During construction of any building with a primary slab floor, an area approximately the footprint of the building to be heated, and > 1 m in depth, is insulated on all 6 sides typically with HDPE closed cell insulation. Pipes are used to transfer solar energy into the insulated area, as well as extracting heat as required on demand. If there is significant internal ground water flow, remedial actions are needed to prevent it.

## Surface and Above Ground Technologies

•   Pit Storage. Lined, shallow dug pits that are filled with gravel and water as the storage medium are used for STES in many Danish district heating systems. Storage pits are covered with a layer of insulation and then soil, and are used for agriculture or other purposes. A system in Marstal, Denmark, includes a pit storage supplied with heat from a field of solar-thermal panels. It is initially providing 20% of the year-round heat for the village and is being expanded to provide twice that. The world's largest pit store (200,000 m³ (7,000,000 cu ft)) was commissioned in Vojens, Denmark, in 2015, and allows solar heat to provide 50% of the annual energy for the world's largest solar-enabled district heating system.

•   Large-scale thermal storage with water. Large scale STES water storage tanks can be built above ground, insulated, and then covered with soil.

•   Horizontal heat exchangers. For small installations, a heat exchanger of corrugated plastic pipe can be shallow-buried in a trench to create a STES.

- Earth-bermed buildings. Stores heat passively in surrounding soil.

- Salt hydrate technology This technology achieves significantly higher storage densities that water-based heat storage.

## Conferences and Organizations

The International Energy Agency's *Energy Conservation through Energy Storage (ECES) Programme* has held triennial global energy conferences since 1981. The conferences originally focused exclusively on STES, but now that those technologies are mature other topics such as phase change materials (PCM) and electrical energy storage are also being covered. Since 1985 each conference has had "stock" (for storage) at the end of its name; e.g. EcoStock, ThermaStock. They are held at various locations around the world. Most recent were InnoStock 2012 (the 12th International Conference on Thermal Energy Storage) in Lleida, Spain and GreenStock 2015 in Beijing. EnerStock 2018 will be held in Adana, Turkey in April 2018.

The IEA-ECES programme continues the work of the earlier *International Council for Thermal Energy Storage* which from 1978 to 1990 had a quarterly newsletter and was initially sponsored by the U.S. Department of Energy. The newsletter was initially called *ATES Newsletter,* and after BTES became a feasible technology it was changed to *STES Newsletter.*

## Use of STES for Small, Passively Heated Buildings

Small passively heated buildings typically use the soil adjoining the building as a low-temperature seasonal heat store that in the annual cycle reaches a maximum temperature similar to average annual air temperature, with the temperature drawn down for heating in colder months. Such systems are a feature of building design, as some simple but significant differences from 'traditional' buildings are necessary. At a depth of about 20 feet (6 m) in the soil, the temperature is naturally stable within a year-round range, if the draw down does not exceed the natural capacity for solar restoration of heat. Such storage systems operate within a narrow range of storage temperatures over the course of a year, as opposed to the other STES systems described above for which large annual temperature differences are intended.

Two basic passive solar building technologies were developed in the US during the 1970s and 1980s. They utilize direct heat conduction to and from thermally isolated, moisture-protected soil as a seasonal storage medium for space heating, with direct conduction as the heat return method. In one method, "passive annual heat storage" (PAHS), the building's windows and other exterior surfaces capture solar heat which is transferred by conduction through the floors, walls, and sometimes the roof, into adjoining thermally buffered soil.

When the interior spaces are cooler than the storage medium, heat is conducted back to the living space. The other method, "annualized geothermal solar" (AGS) uses a separate solar collector to capture heat. The collected heat is delivered to a storage device (soil, gravel bed or water tank) either passively by the convection of the heat transfer medium (e.g. air or water) or actively by pumping it. This method is usually implemented with a capacity designed for six months of heating.

A number of examples of the use of solar thermal storage from across the world include: Suffolk One a college in East Anglia, England, that uses a thermal collector of pipe buried in the bus turning area to collect solar energy that is then stored in 18 boreholes each 100 metres (330 ft) deep for use in winter heating. Drake Landing Solar Community in Canada uses solar thermal collectors on the garage roofs of 52 homes, which is then stored in an array of 35 metres (115 ft) deep boreholes. The ground can reach temperatures in excess of 70 °C which is then used to heat the houses passively. The scheme has been running successfully since 2007. In Brædstrup, Denmark, some 8,000 square metres (86,000 sq ft) of solar thermal collectors are used to collect some 4,000,000 kWh/year similarly stored in an array of 50 metres (160 ft) deep boreholes.

## Liquid Engineering

Architect Matyas Gutai obtained an EU grant to construct a house in Hungary which uses extensive water filled wall panels as heat collectors and reservoirs with underground heat storage water tanks. The design uses microprocessor control.

## Small Buildings with Internal STES Water Tanks

A number of homes and small apartment buildings have demonstrated combining a large internal water tank for heat storage with roof-mounted solar-thermal collectors. Storage temperatures of 90 °C (194 °F) are sufficient to supply both domestic hot water and space heating. The first such house was MIT Solar House #1, in 1939. An eight-unit apartment building in Oberburg, Switzerland was built in 1989, with three tanks storing a total of 118 m$^3$ (4,167 cubic feet) that store more heat than the building requires. Since 2011, that design is now being replicated in new buildings.

In Berlin, the "Zero Heating Energy House", was built in 1997 in as part of the IEA Task 13 low energy housing demonstration project. It stores water at temperatures up to 90 °C (194 °F) inside a 20 m$^3$ (706 cubic feet) tank in the basement.

A similar example was built in Ireland in 2009, as a prototype. The *solar seasonal store* consists of a 23 m$^3$ (812 cu ft) tank, filled with water, which was installed in the ground, heavily insulated all around, to store heat from evacuated solar tubes during the year. The system was installed as an experiment to heat the *world's first standardized pre-fabricated passive house* in Galway, Ireland. The aim was to find out if this heat would be sufficient to eliminate the need for any electricity in the already highly efficient home during the winter months.

## Use of STES in Greenhouses

STES is also used extensively for the heating of greenhouses. ATES is the kind of storage commonly in use for this application. In summer, the greenhouse is cooled with ground water, pumped from the "cold well" in the aquifer. The water is heated in the process, and is returned to the "warm well" in the aquifer. When the greenhouse needs heat, such as to extend the growing season, water is withdrawn from the warm well, becomes chilled while serving its heating function, and is returned to the cold well. This is a very efficient system of free cooling, which uses only circulation pumps and no heat pumps.

## Thermal Battery

A thermal energy battery is a physical structure used for the purpose of storing and releasing thermal energy. Such a thermal battery (a.k.a. TBat) allows energy available at one time to be temporarily stored and then released at another time. The basic principles involved in a thermal battery occur at the atomic level of matter, with energy being added to or taken from either a solid mass or a liquid volume which causes the substance's temperature to change. Some thermal batteries also involve causing a substance to transition thermally through a phase transition which causes even more energy to be stored and released due to the delta enthalpy of fusion or delta enthalpy of vaporization.

## History of Thermal Batteries

Thermal batteries are very common, and include such familiar items as a hot water bottle. Early examples of thermal batteries would include stone and mud cook stoves, rocks placed in fires, and kilns. While stoves and kilns are ovens, then are also thermal storage systems that depend on heat being retained for an extended period of time.

## Types of Thermal Batteries

Thermal batteries generally fall into 4 categories:

- GHEX thermal batteries

- Encapsulated thermal batteries

- Phase change thermal batteries

- Other thermal batteries

These 4 types of batteries are each unique in their form and application, although fundamentally all are for the storage and retrieval of thermal energy. They also differ in method and density of heat storage. A description of each type of thermal battery follows.

## GHEx Thermal Battery - Unencapsulated

A ground heat exchanger (GHEX) is an area of the earth that is utilized as an annual cycle thermal battery. These thermal batteries are areas of the earth into which pipes have been placed in order to transfer thermal energy; they are "unencapsulated" in the sense that the target area is not insulated from the rest of the surrounding earth. Energy is added to the GHEX by running a higher temperature fluid through the pipes and thus raising the temperature of the local earth. Energy can also be taken from the GHEX by running a lower temperature fluid through those same pipes.

GHEX thermal batteries are implemented in two forms. The picture above depicts what is known as a "horizontal" GHEX where trenching is used to place an amount of pipe in a closed loop in the ground. They are also formed by drilling boreholes into the ground, either vertically or horizontally, and then the pipes are inserted in the form of a closed-loop with a "u-bend" fitting on the far end of the loop. These drilled GHEX thermal batteries are also sometimes called "borehole thermal energy storage systems".

Heat energy can be added to or removed from a GHEX thermal battery at any point in time. However, they are most often used on an annual cycle where energy is extracted from a building during the summer season to cool a building and added to the GHEX, and then that same energy is later extracted from the GHEX in the winter season to heat the building. This annual cycle of energy addition and subtraction is highly predictable based on energy modeling of the building served. A thermal battery used in this mode is a renewable energy source as the energy extracted in the winter will be restored to the GHEX the next summer in a continually repeating cycle. This type is solar powered because it is the heat from the sun in the summer that is removed from a building and stored in the ground for use in the next winter season for heating.

## Phase Change Thermal Battery

Phase change materials used for thermal storage are capable of storing and releasing significant thermal capacity at the temperature that they change phase. These materials are chosen based on specific applications because there is a wide range of temperatures that may be useful in different applications and a wide range of materials that change phase at different temperatures. These materials include salts and waxes that are specifically engineered for the applications they serve. In addition to manufactured materials, water is a phase change material. The latent heat of water is 334 joules/gram. The phase change of water occurs at 0°C (32°F).

Some applications use the thermal capacity of water or ice as cold storage; others use it as heat storage. It can serve either application; ice can be melted to store heat then refrozen to warm an environment which is below freezing (putting liquid water at 0°C in such an environment warms it much more than the same mass of ice at the same temperature, because the latent heat of freezing is extracted from it, which is why the phase change is relevant), or water can be frozen to "store cold" then melted to make an environment above freezing colder (and again, a given mass of ice at 0°C will provide more cooling than the same mass of water at the same temperature).

The advantage of using a phase change in this way is that a given mass of material can absorb a large quantity of energy without its temperature changing. Hence a thermal battery that uses a phase change can be made lighter, or more energy can be put into it without raising the internal temperature unacceptably.

## Encapsulated Thermal Battery

An encapsulated thermal battery is physically similar to a phase change thermal battery in that it is a confined amount of physical material which is thermally heated or cooled to store or extract energy. However, in a non-phase change encapsulated thermal battery the temperature of the substance is changed without inducing a phase change. Since a phase change is not needed many more materials are available for use in an encapsulated thermal battery.

One of the key properties of an encapsulated thermal battery is its volumetric heat capacity (VHC), also termed volume-specific heat capacity. Typical substances used for these thermal batteries include water, concrete, and wet sand.

An example of an encapsulated thermal battery is a residential water heater with a storage tank. This thermal battery is usually slowly charged over a period of about 30–60 minutes for rapid use when needed (e.g., 10–15 minutes). Many utilities, understanding the "thermal battery" nature of water heaters, have begun using them to absorb excess renewable energy power when available for later use by the homeowner. According to the above cited article, "net savings to the electricity system as a whole could be $200 per year per heater – some of which may be passed on to its owner".

## Other Thermal Batteries

There are some other items that have historically been termed "thermal batteries". In this group is the molten salt battery which is a device for generating electricity. Other examples include the heat packs that skiers use for keeping hands and feet warm. These are a chemical battery which when activated (with air in this case) will produce heat. Other related chemical thermal batteries exist for producing cold generally used for sport injuries.

The one common principle of these other thermal batteries is that the reaction involved is generally not reversible. Thus, these batteries are not used for storing and retrieving heat energy.

## Ice Storage Air Conditioning

Ice storage air conditioning is the process of using ice for thermal energy storage. This is practical because of water's large heat of fusion: one metric ton of water (one cubic metre) can store 334 megajoules (MJ) (317,000 BTU) of energy, equivalent to 93 kWh (26.4 ton-hours).

Ice was originally obtained from mountains or cut from frozen lakes and transported to cities for use as a coolant. The original definition of a "ton of cooling capacity" (heat flow) was the heat needed to melt one ton of ice in a 24-hour period. This heat flow is what one would expect in a 3,000-square-foot (280 m²) house in Boston in the summer. This definition has since been replaced by less archaic units: one ton HVAC capacity is equal to 12,000 BTU per hour. A small storage facility can hold enough ice to cool a large building from one day to one week, whether that ice is produced by anhydrous ammonia chillers or hauled in by horse-drawn carts.

Ground freezing can also be utilized; this may be done in ice form where the ground is saturated. Systems will also work with pure rock. Wherever ice forms, the ice formation's heat of fusion is not used, as the ice remains solid throughout the process. The method based on ground freezing is widely used for mining and tunneling to solidify unstable ground during excavations. The ground is frozen using bore holes with concentric pipes that carry brine from a chiller at the surface. Cold is extracted in a similar way using brine and used in the same way as for conventional ice storage, normally with a brine-to-liquid heat exchanger, to bring the working temperatures up to usable levels at higher volumes. The frozen ground can stay cold for months or longer, allowing cold storage for extended periods at negligible structure cost.

Replacing existing air conditioning systems with ice storage offers a cost-effective energy storage method, enabling surplus wind energy and other such intermittent energy sources to be stored for use in chilling at a later time, possibly months later.

## Air Conditioning

The most widely used form of this technology can be found in campus-wide air conditioning or chilled water systems of large buildings. Air conditioning systems, especially in commercial buildings, are the biggest contributors to peak electrical loads seen on hot summer days in various countries. In this application, a standard chiller runs at night to produce an ice pile. Water then circulates through the pile during the day to produce chilled water that would normally be the chiller's daytime output.

A partial storage system minimizes capital investment by running the chillers nearly 24 hours a day. At night, they produce ice for storage and during the day they chill water for the air conditioning system. Water circulating through the melting ice augments their production. Such a system usually runs in ice-making mode for 16 to 18 hours a day and in ice-melting mode for six hours a day. Capital expenditures are minimized because the chillers can be just 40 - 50% of the size needed for a conventional design. Ice storage sufficient to store half a day's rejected heat is usually adequate.

A full storage system minimizes the cost of energy to run that system by entirely shutting off the chillers during peak load hours. The capital cost is higher, as such a system requires somewhat larger chillers than those from a partial storage system, and a larger ice storage system. Ice storage systems are inexpensive enough that full storage systems are often competitive with conventional air conditioning designs.

The air conditioning chillers' efficiency is measured by their coefficient of performance (COP). In theory, thermal storage systems could make chillers more efficient because heat is discharged into colder nighttime air rather than warmer daytime air. In practice, heat loss overpowers this advantage, since it melts the ice.

Air conditioning thermal storage has been shown to be somewhat beneficial in society. The fuel used at night to produce electricity is a domestic resource in most countries, so less imported fuel is used. Also, studies show that this process significantly reduces the emissions associated with producing the power for air conditioners, since in the evening, inefficient "peaker" plants are replaced by low-emission base load facilities. The plants that produce this power often work more efficiently than the gas turbines that provide peaking power during the day. As well, since the load factor on the plants is higher, fewer plants are needed to service the load.

A new twist on this technology uses ice as a condensing medium for the refrigerant. In this case, regular refrigerant is pumped to coils where it is used. Rather than needing a compressor to convert it back into a liquid, however, the low temperature of ice is used to chill the refrigerant back into a liquid. This type of system allows existing refrigerant-based HVAC equipment to be converted to Thermal Energy Storage systems, something that could not previously be easily done with chill water technology. In addition, unlike water-cooled chill water systems that do not experience a tremendous difference in efficiency from day to night, this new class of equipment typically displaces daytime operation of air-cooled condensing units. In areas where there is a significant difference between peak day time temperatures and off peak temperatures, this type of unit is typically more energy efficient than the equipment that it replaces.

## Combustion Gas Turbine Air Inlet Cooling

Thermal energy storage is also used for combustion gas turbine air inlet cooling. Instead of shifting electrical demand to the night, this technique shifts generation capacity to the day. To generate ice at night, the turbine is often mechanically connected to a large chiller's compressor. During peak daytime loads, water is circulated between the ice pile and a heat exchanger in front of the turbine air intake, cooling the intake air to near freezing temperatures. Since the air is colder, the turbine can compress more air with a given amount of compressor power. Typically, both the generated electrical power and turbine efficiency rise when the inlet cooling system is activated. This system is similar to the compressed air energy storage system.

## Layered Charge Storage

Two long-term solar storage units installed in solar-powered apartment buildings

Layered or stratified charge storage is hot water storage tank, typically for solar thermal energy. The warmest storage layer is the top storage cylinder and below this there are colder storage layers through natural layering. The water is fed into different storage levels, depending on the available feed temperature and current temperature layering. The feed takes place via a vertical line via valves, in each case the feed water is fed into the storage layer with the corresponding water temperature. This is achieved by means of self-acting valves, feeding the respective temperature or water density into the temperature layer. The advantage is that there is no mixing of the storage temperature. Swirling of the water is therefore to be avoided as far as possible in stratified storage tanks because this could lead to intermingling of storage layers. This is reduced by means of measures such as limiting flow velocity and baffle plates at the inlets.

Heating is therefore only necessary later, since warm water can be taken from above for longer, as opposed to mixed storage. In addition, water can be drawn with almost the same temperature very quickly after filling with hot water, since not all of the storage tank has to be heated. This optimises energy efficiency.

At the moment, they are generally more expensive than combined storage tank, but the storage volume can be a bit smaller and they are considered more modern because of better energy consumption.

In addition to storage for, for example, single-family houses with a few hundred to over one thousand litres, they can also be found in larger and significantly larger forms, for example, as long-term thermal storage tanks with a few thousand litres of storage volume.

Areas of application

- For solar thermal systems, to bridge periods of no sun

- For wood heating, since wood heating is otherwise difficult to control

- If a heat pump is installed in order to maintain blocking and standstill times

- For combined heat and power plants, to use them more efficiently

- For connecting different heating systems

## Oil Terminal

An oil depot in Kowloon, Hong Kong around the mid-1980s.
The depot was redeveloped into a residential area Laguna City in the late 80s and early 90s.

Tank farm at McMurdo Station, Antarctica

An oil depot (sometimes called a tank farm, installation or oil terminal) is an industrial facility for the storage of oil and/or petrochemical products and from which these products are usually transported to end users or further storage facilities. An oil depot typically has tankage, either above ground or underground, and gantries (framework) for the discharge of products into road tankers or other vehicles (such as barges) or pipelines.

Oil depots are usually situated close to oil refineries or in locations where marine tankers containing products can discharge their cargo. Some depots are attached to pipelines from which they draw their supplies and depots can also be fed by rail, by barge and by road tanker (sometimes known as "bridging").

Most oil depots have road tankers operating from their grounds and these vehicles transport products to petrol stations or other users.

An oil depot is a comparatively unsophisticated facility in that (in most cases) there is no processing or other transformation on site. The products which reach the depot (from a refinery) are in their final form suitable for delivery to customers. In some cases additives may be injected into products in tanks, but there is usually no manufacturing plant on site. Modern depots comprise the same types of tankage, pipelines and gantries as those in the past and although there is a greater degree of automation on site, there have been few significant changes in depot operational activities over time.

## Health, Safety and Environment

One of the key imperatives is Health, Safety and Environment (HSE) and the operators of a depot must ensure that products are safely stored and handled. There must be no leakages (etc.) which could damage the soil or the water table.

Massive fire at Buncefield Oil Depot, UK December 2005

Fire protection is a primary consideration, especially for the more flammable products such as petrol (gasoline) and Aviation Fuel.

## Ownership

The ownership of oil depots falls into three main categories:

- Single oil company ownership. When one company owns and operates a depot on its own behalf.

- Joint or consortium ownership, where two or more companies own a depot together and share its operating costs.

- Independent ownership, where a depot is owned not by an oil company but by a separate business which charges oil companies (and others) a fee to store and handle products. The Royal Vopak from the Netherlands is the largest independent terminal operator with 80 terminals in 30 countries.

In all cases the owners may also provide "hospitality" or "pick up rights" at the facility to other companies.

## Airports

Aircraft refueller at Vancouver airport

Most airports also have their own dedicated oil depots (usually called "fuel farms") where aviation fuel (Jet A or 100LL) is stored prior to being discharged into aircraft fuel tanks. Fuel is transported from the depot to the aircraft either by road tanker or via a hydrant system.

## Japan

The world's third largest oil consumer had national reserves of 113 days of oil demand under the government's storage and 85 days held by the private sector at the end of December 2010. In this respect, the total oil stored in Japan in December stood at 587.4 million barrels. Japan requires the private sector to hold 70 days as oil reserves, but is making the period shorter by three days to 67 days. As such it will allow oil companies to release 8.9 million barrels of crude oil from mandatory stockpiles.

## Floating Production Storage and Offloading

FPSO *Mystras* at work off the shore of Nigeria

FPSO *Crystal Ocean* moored at the Port of Melbourne

The circular FPSO *Sevan Voyageur* moored at Nymo yard at Eydehavn, Norway

A floating production storage and offloading (FPSO) unit is a floating vessel used by the offshore oil and gas industry for the production and processing of hydrocarbons, and for the storage of oil. A FPSO vessel is designed to receive hydrocarbons produced by itself or from nearby platforms or subsea template, process them, and store oil until it can be offloaded onto a tanker or, less frequently, transported through a pipeline. FPSOs are preferred in frontier offshore regions as they are easy to install, and do not require a local pipeline infrastructure to export oil. FPSOs can be a conversion of an oil tanker or can be a vessel built specially for the application. A vessel used only to store oil (without processing it) is referred to as a floating storage and offloading (FSO) vessel.

Recent developments in LNG industry require relocation of conventional LNG processing trains into the sea to unlock remote, smaller gas fields that would not be economical to develop otherwise, reduce capital expenses, and impact to environment. Emerging new type of FLNG facilities will be used. Unlike FPSO's apart of gas production, storage and offloading, they will also allow full scale deep processing, same as onshore LNG plant has to offer but squeezed to 25% of its footprint. First 3 FLNG's are under construction (as at 2016): Prelude FLNG (Shell), PFLNG1 and PFLNG2 (Petronas).

## History

Oil has been produced from offshore locations since the late 1940s. Originally, all oil platforms sat on the seabed, but as exploration moved to deeper waters and more distant locations in the 1970s, floating production systems came to be used.

The first oil FPSO was the *Shell Castellon*, built in Spain in 1977. Today, over 270 vessels are deployed worldwide as oil FPSOs.

On July 29, 2009, Shell and Samsung announced an agreement to build up to 10 LNG FPSOs, at same Samsung Yard Flex LNG appeared to construct smaller units.

On May 20, 2011, Royal Dutch Shell announced the planned development of a floating liquefied natural gas facility (FLNG), called Prelude with 488 m long and 74 m wide, which is to be situated 200 km off the coast of Western Australia and is due for completion in around 2016, the largest vessel man-made ever. Royal Dutch Shell (2013), LNG FPSO (Liquefied Natural Gas Floating production Storage and Offloading), Samsung Heavy Industries at a cost of $12 Billion.

In June 2012, Petronas made a contract of procurement engineering, construction, installation and commissioning, a project with the Technip and DSME consortium. The unit is destined for the Kanowit gas field off Sarawak, Malaysia. It is expected to be the World's First Floating Liquefaction Unit in operation when completed in 2015.

At the opposite (discharge and regasification) end of the LNG chain, the first ever conversion of an LNG carrier, Golar LNG owned Moss type LNG carrier into an LNG floating storage and regasification unit was carried out in 2007 by Keppel shipyard in Singapore.

## Mechanisms

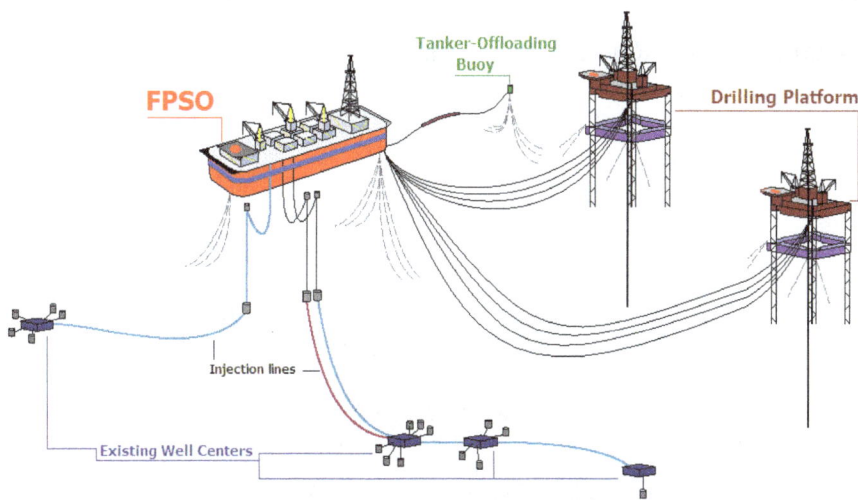

FPSO diagram

Oil produced from offshore production platforms can be transported to the mainland either by pipeline or by tanker. When a tanker is chosen to transport the oil, it is necessary to accumulate oil in some form of storage tank, such that the oil tanker is not continuously occupied during oil production, and is only needed once sufficient oil has been produced to fill the tanker.

## Advantages

Floating production, storage and offloading vessels are particularly effective in remote or deep water locations, where seabed pipelines are not cost effective. FPSOs eliminate the need to lay expensive long-distance pipelines from the processing facility to an onshore terminal. This can provide an economically attractive solution for smaller *oil fields*, which can be exhausted in a few years and do not justify the expense of installing a pipeline. Furthermore, once the field is depleted, the FPSO can be moved to a new location.

## Types

A Floating Storage and Offloading unit (FSO) is essentially a simplified FPSO, without the capability for oil or gas processing. Most FSOs are converted *single hull* supertankers. An example is *Knock Nevis*, ex *Seawise Giant*, which for many years was the world's largest ship. It was converted into an FSO for use offshore use before being scrapped.

At the other end of the LNG logistics chain, where the natural gas is brought back to ambient temperature and pressure, specially modified ships may also be used as floating storage and regasification units (FSRUs). A LNG floating storage and regasification unit receives liquefied natural gas (LNG) from offloading LNG carriers, and the onboard regasification system provides natural gas exported to shore through risers and pipelines.

1.      FSO, *F*loating *S*torage and *O*ffloading

2.      FPSO, *F*loating *P*roduction, *S*torage and *O*ffloading

3.      FDPSO, *F*loating, *D*rilling and *P*roduction, *S*torage and *O*ffloading

4.      FSRU, *F*loating *S*torage *R*egasification *U*nit.

## Vessels

## Records

FPSO *Firenze* moored at Hellenic Shipyards, 2007

The FPSO operating in the deepest waters is the FPSO *BW Pioneer*, built and operated by BW Offshore on behalf of Petrobras Americas INC. The FPSO is moored at a depth of 2,600 m in Block 249 Walker Ridge in the US Gulf of Mexico and is rated for 80,000 bbl/d (13,000 m³/d). The EPCI contract was awarded in October 2007 and production started in early 2012. The FPSO conversion was carried out at *MMHE Shipyard Pasir Gudang* in Malaysia, while the topsides were fabricated in modules at various international vendor locations. The FPSO has a disconnectable turret (APL). The vessel can disconnect in advance of hurricanes and reconnect with minimal down time. A contract for an FPSO to operate in even deeper waters (2,900 m) for Shell's Stones field in the US Gulf of Mexico was awarded to SBM Offshore in July 2013.

One of the world's largest FPSO is the *Kizomba A*, with a storage capacity of 2.2 million barrels (350,000 m³). Built at a cost of over US$ 800 million by Hyundai Heavy Industries in Ulsan, Korea, it is operated by Esso Exploration Angola (ExxonMobil). Located in 1200 meters (3,940 ft) of water at Deep water block 200 statute miles (320 km) offshore in the Atlantic Ocean from Angola, Central Africa, it weighs 81,000 tonnes and is 285 meters long, 63 meters wide, and 32 meters high (935 ft by 207 ft (63 m) by 105 ft).

The first FSO in the Gulf of Mexico, The FSO *Ta'Kuntah*, has been in operation since August 1998. The FSO, owned and operated by MODEC, is under a service agreement with *PEMEX* Exploration and Production. The vessel was installed as part of the Cantarell Field Development. The field is located in the Bay of Campeche, offshore Mexico's Yucatán peninsula. It is a converted ULCC tanker with a SOFEC external turret mooring system, two flexible risers connected in a lazy-S configuration between the turret and a pipeline end manifold (PLEM) on the seabed, and a unique offloading system. The FSO is designed to handle 800,000 bbl/d (130,000 m³/d) with no allowance for downtime.

The Skarv FPSO, developed and engineered by *Aker Solutions* for *BP Norge*, is one of the most advanced and largest FPSO deployed in the Norwegian Sea, offshore Mid Norway. *Skarv* is a gas condensate and oil field development. The development ties in five sub-sea templates, and the FPSO has capacity to include several smaller wells nearby in the future. The process plant on the vessel can handle about 19,000,000 cubic metres per day (670,000,000 cu ft/d) of gas and 13,500 cubic metres per day (480,000 cu ft/d) of oil. An 80 km gas export pipe ties into Åsgard transport system. *Aker Solutions* (formerly Aker Kvaerner) developed the front-end design for the floating production facility as well as the overall system design for the field and preparation for procurement and project management of the total field development. The hull is an Aker Solutions proprietary "Tentech975" design. BP also selected Aker Solutions to perform the detail engineering, procurement and construction management assistance (EPcma) for the Skarv field development. The *EPcma* contract covers detail engineering and procurement work for the FPSO topsides as well as construction management assistance to BP including hull and topside facilities. The production started in field on August 2011. BP awarded the contract for fabrication of the *Skarv* FPSO hull to *Samsung Heavy Industries* in South Korea and the Turret contract to SBM. The FPSO has a length of 292 m, beam of 50.6 m and is 29 m deep, accommodates about 100 people in single cabins. The hull is delivered in January 2010.

## Central Oil Storage

Central oil storage (COS), or central storage, is the term used for a communal heating system that began to be utilized in the middle of the twentieth century.

The term also applies to industrial, plant and agricultural applications, all of which may not be communal in nature.

## Household Applications

The concept involved using oil (usually kerosene, but sometimes gas oil), in the way that natural gas is used today – it being fed from a central source and metered into individual dwellings on a housing estate. The concept was utilized by major oil companies such as Shell and BP. Each household was obliged to use their oil from their tank, unlike the situation where a consumer owns their own tank obtains their own oil.

The book *Introduction to Architectural Science* states about liquid fuel storage tanks, "often in a housing development a central storage tank is installed (usually underground) which will be filled by an oil company", and that a supply of liquid fuel is piped to individual apartments or houses from the central storage tank.

## Logistics

Oil would be delivered by road tanker and discharged into a tank capable of holding, for example, 5000 gallons. This oil would then be piped to each home, initially through a master pipeline, which then subdivided underground with a branch leading into each property. Each property was provided with a meter located on an outside wall which was read whenever necessary.

A 5,000-gallon tank would usually be installed on an estate numbering up to 50 or 60 properties, although they ranged from a low of about 12 up to a high of in the thousands. The majority of tanks were situated above ground in an elevated position. This meant the oil could flow into each house under the influence of gravity. Other tanks were situated underground and were equipped with an electric pump, usually feeding oil to a small 'header' tank, again in an elevated position, to allow gravity to then distribute the oil.

The level of oil in the tank would be checked regularly and new supplies ordered when necessary. One potential problem with this system is that it was possible, through negligence, for the tank to run dry, resulting in a number of properties being left with no heat.

From the early 1980s, their owners began to close down COS sites. The significant increases in the price of oil had led many customers to convert to gas, solid fuel, or even to install their own oil tank. With fewer and fewer users per site, and maintenance costs remaining the same, the oil companies went through a closure programme, resulting in few sites still being operational in the new millennium.

As at early 2008, the ownership of many sites is unclear. Sites have been sold through a number of companies which have then been closed down. This presents problems for householders and local authorities alike, for when a site experiences a problem – such as an unsafe bund wall, or even a leak – there is no one to pursue to correct matters.

## Industrial Applications

Central oil storage is also performed at industrial and plant locations and operations. Central oil storage in industrial applications may be utilized in part to conserve oil, because oil barrels may leak oil.

## References

- He, Teng; Pei, Qijun; Chen, Ping (2015-09-01). "Liquid organic hydrogen carriers". Journal of Energy Chemistry. 24 (5): 587–594. doi:10.1016/j.jechem.2015.08.007

- "Siemens project to test heated rocks for large-scale, low-cost thermal energy storage". Utility Dive. 12 October 2016. Retrieved 15 October 2016

- Wang, Bo; Goodman, D. Wayne; Froment, Gilbert F. (2008-01-25). "Kinetic modeling of pure hydrogen production from decalin". Journal of Catalysis. 253 (2): 229–238. doi:10.1016/j.jcat.2007.11.012

- Grasemann, Martin; Laurenczy, Gábor. "Formic acid as a hydrogen source – recent developments and future trends". pubs.rsc.org. doi:10.1039/C2EE21928J. Retrieved 2015-11-04

- Ehrlich, Robert (2013). "Thermal storage". Renewable Energy: A First Course. CRC Press. p. 375. ISBN 978-1-4398-6115-8

- "Thermal capacitors made from Miscibility Gap Alloys (MGAs) (PDF Download Available)". ResearchGate. Retrieved 2017-02-27

- Zhevago, N.K.; Denisov, E.I.; Glebov, V.I. (2010). "Experimental investigation of hydrogen storage in capillary arrays". International Journal of Hydrogen Energy. 35: 169–175. doi:10.1016/j.ijhydene.2009.10.011

- Welch, G. C.; Juan, R. R. S.; Masuda, J. D.; Stephan, D. W. (2006). "Reversible, Metal-Free Hydrogen Activation". Science. 314 (5802): 1124–6. PMID 17110572. doi:10.1126/science.1134230

- Wong, Bill (June 28, 2011), "Drake Landing Solar Community" (PDF), Drake Landing Solar Community, IDEA/CDEA District Energy/CHP 2011 Conference, Toronto, pp. 1–30, retrieved 21 April 2013

- Hestnes, A.; Hastings, R. (eds) (2003). Solar Energy Houses: Strategies, Technologies, Examples. pp. 109-114. ISBN 1-902916-43-3

- Brünig, Thorge; Krekic, Kristijan; Bruhn, Clemens; Pietschnig, Rudolf (2016). "Calorimetric Studies and Structural Aspects of Ionic Liquids in Designing Sorption Materials for Thermal Energy Storage". Chemistry European Journal. 22: 16200–16212. doi:10.1002/chem.201602723

- Zhevago, N.K.; Glebov, V.I. (2007). "Hydrogen storage in capillary arrays". Energy Conversion and Management. 48 (5): 1554–1559. doi:10.1016/j.enconman.2006.11.017

- Kelly-Detwiler, Peter. "Ice Storage: A Cost-Efficient Way To Cool Commercial Buildings While Optimizing the Power Grid". Retrieved 20 June 2017

- Kolpak, Alexie M.; Grossman, Jeffrey C. (2011). "Azobenzene-Functionalized Carbon Nanotubes As High-Energy Density Solar Thermal Fuels". Nano Letters. 11 (8): 3156–62. Bibcode:2011NanoL..11.3156K. PMID 21688811. doi:10.1021/nl201357n

# Diverse Aspects of Energy Storage

Power-to-x refers to the excess electricity that is used by energy storage and conversion pathways when energy from renewable sources fluctuates. Energy carrier, Leyden jar, dry-seal Wiggins gasholder, etc. are some of the other topics that have been explored. Energy storage is best understood in confluence with the major topics listed in the following chapter.

## Power-to-X

Power-to-X (also P2X and P2Y) has two meanings. First, power-to-X refers to a number of electricity conversion, energy storage, and reconversion pathways that utilize surplus electric power, typically during periods where fluctuating renewable energy generation exceeds load. Second, power-to-X refers to conversion technologies that allow for the decoupling of power from the electricity sector for use in other sectors (such as transport or chemicals), possibly using power that has been provided by additional investments in generation. The term power-to-x is widely used in Germany and may have originated there.

The X in the terminology can refer to one of the following: power-to-ammonia, power-to-chemicals, power-to-fuel, power-to-gas, power-to-heat, power-to-hydrogen, power-to-liquid, power-to-methane, power-to-mobility, power-to-power, and power-to-syngas.

Collectively power-to-X schemes which use surplus power fall under the heading of flexibility measures and are particularly useful in energy systems with high shares of renewable generation and/or with strong decarbonization targets. A large number of pathways and technologies are encompassed by the term. In 2016 the German government funded a €30 million first-phase research project into power-to-X options.

### Electricity Storage Concepts

Surplus electric power can be converted to other forms of energy for storage and reconversion. Direct current electrolysis (efficiency 80–85% at best) can be used to produce hydrogen which can, in turn, be converted to methane ($CH_4$) via methanation. These fuels can be stored and used to produce electricity again, hours to months later. Reconversion technologies include gas turbines, CCGT plant, and fuel cells. Power-to-power refers to the round-trip reconversion efficiency. For hydrogen storage, the round-trip efficiency remains limited at 35–50%. Electrolysis is expensive and power-to-gas processes need substantial full-load hours (say 30%) to be economic. Grid-dedicated battery storage is not normally considered a power-to-X concept.

### Sector Coupling Concepts

Hydrogen and methane can also be used as downstream fuels, feed into the natural gas grid, or used to make or synthetic fuel. Alternatively they can be used as a chemical feedstock, as can ammonia ($NH_3$).

Power-to-heat involves contributing to the heat sector, either by resistance heating or via a heat pump. Resistance heaters have unity efficiency, and the corresponding coefficient of performance (COP) of heat pumps is 2–5. Back-up immersion heating of both domestic hot water and district heating offers a cheap way of using surplus renewable energy and will often displace carbon-intensive fossil fuels for the task. Large-scale heat pumps in district heating systems with thermal energy storage are an especially attractive option for power-to-heat: they offer exceptionally high efficiency for balancing excess wind and solar power, and they can be profitable investments.

Power-to-mobility refers to the charging of battery electric vehicles (EV). Given the expected uptake of EVs, dedicated dispatch will be required. As vehicles are idle for most of the time, shifting the charging time can offer considerable flexibility: the charging window is a relatively long 8–12 hours, whereas the charging duration is around 90 minutes. The EV batteries can also be discharged to the grid to make them work as electricity storage devices, but this causes additional wear to the battery.

Heat pumps with hot water storage and electric vehicles have been found to have higher potential on reduction of $CO_2$ emissions and fossil fuel use than several other power-to-X or electricity storage schemes for using surplus wind and solar power.

## Energy Storage as a Service (ESaaS)

Energy Storage as a Service (ESaaS) allows a facility to benefit from the advantages of an energy storage system by entering into a service agreement without purchasing the system. Energy storage systems provide a range of services to generate revenue, create savings, and improve electricity resiliency. The operation of the ESaaS system is a unique combination of an advanced battery storage system, an energy management system, and a service contract which can deliver value to a business by providing reliable power more economically.

### History

The term ESaaS was developed and trademarked by Constant Power Inc., a Toronto-based company, in 2016. The service has been designed to work in the North American open electricity markets. Notable other companies offering Energy Storage-as-a-Service include AES Corporation and STEM.

### Components

ESaaS is the combination of an energy storage system, a control and monitoring system, and a service contract.

The most common energy storage systems used for ESaaS are lithium-ion or flow batteries due to their compact size, non-invasive installation, high efficiencies, and fast reaction times but other storage mediums may be used such as compressed air, flywheels, or pumped hydro. The batteries are sized based on the facility's needs and is paired with a power inverter to convert the DC power to AC power in order to connect directly to the facility's electricity supply.

ESaaS systems are remotely monitored and controlled by the ESaaS operator using a Supervisory Control and Data Acquisition (SCADA) system. The SCADA communicates with the facility's Energy Management System (EMS), Power Conversion System (PCS), and Battery Management System (BMS). The ESaaS operator is responsible for ensuring the ESaaS system is monitoring and responding to the facility's needs as well as overriding commands to participate in regional incentive programs such as coincident peak management and demand response programs in real time.

The facility benefiting from the ESaaS system is linked to the ESaaS system operator through a service contract. The contract specifies the length of the service term, payment structure, and list of services the facility wishes to participate in.

## Services

ESaaS is used to perform a variety of services including:

- Coincident Peak Management

  During times of high regional demand, Independent Service Operators (ISOs)/Regional Transmission Organizations (RTOs) offer incentives for facilities to reduce or curtail their load. ESaaS allows a facility to isolate or offset their load during these high regional demand periods to decrease demand from the electricity grid to benefit from the incentives. The system is designed to work in conjunction or independent of facility curtailment.

- Demand Response

  ISOs/RTOs offer facilities payment for curtailing their energy demand when dispatched by the grid operator. ESaaS allows facilities to participate in these programs by off-setting all or a portion of a facility load during a demand response occurrence. A facility can benefit from the incentive without interrupting their facility operation.

- Power Factor Correction

  During charging and discharging, active and reactive power may be balanced prior to supplying a facility. By balancing the amount of active and reactive power to a facility, the power factor and resulting facility electrical efficiency may be improved. This improvement may reduce a facility's monthly peak demand charge.

- Power Quality

  ESaaS actively monitors electricity supply to a facility. In times of intermittent power supply, ESaaS acts as an uninterruptible power supply (UPS) to ensure uninterrupted, reliable power supply to eliminate unexpected fluctuations. Fluctuating and intermittent power affects equipment operation which may cause costly delays and defects in production.

- Back-up Power

  If the electricity grid experiences a power outage, ESaaS offers a back-up power service to continue powering all or a portion of a facility's electricity demand. Depending on the size of the ESaaS installation, ESaaS may maintain facility operation for the duration of a grid failure.

- Peak Shaving

  ESaaS actively monitors a facility's energy profile to normalize the electricity draw from the electricity grid. The ESaaS system stores energy when the facility demand is lower than average and discharges the stored energy when the facility demand is higher than average. The result is a steady draw of electricity from the electricity grid and a lower monthly peak demand charge.

- Energy Arbitrage

  ESaaS actively monitors local electricity spot prices to store energy when the price is low to be utilized when electricity prices are high. This is commonly referred to as arbitrage. The net different in price results in cost savings.

- Market Ancillary Services

  ESaaS enables facilities to participate in the local ISO/RTO markets to provide services such as frequency regulation, operating reserve, and dispatchable generation. By participating in the local market, facilities can generate revenue through the ESaaS contract.

- Transmission Support

  ESaaS may provide services to ease congestion and constraint on electricity transmission networks by storing energy during heavy transmission periods to be released during less congested periods. The use of this service can prolong the life of infrastructure and defers system upgrades.

## Markets served

ESaaS primarily benefits large energy consumers with an average demand of over 500 kW, although, the service may benefit smaller facilities depending on regional incentives. Current early adopters of ESaaS are manufacturers (chemical, electrical, lighting, metal, petrochemical, plastics), commercial (retail, large offices, medium offices, multi-residential, supermarkets), public facilities (colleges, universities, hotels, hospitality, schools), and resources (oil & extraction, pulp & paper, metals & ore, food processing, greenhouses).

## Benefits

## System Benefactor does not Require Installation Capital

To participate in an ESaaS service, the installation system benefactor does not require any capital outlay. Upon installing an ESaaS service, the facility sees immediate savings and/or revenue generation. Initial capital is often a hurdle for facilities to adopt an energy storage system since in most cases, the payback period of an energy storage system is 5–10 years.

### System Operated by a Third-party System Operator

ESaaS is a contracted service that is automatically controlled by a third party. This eliminates responsibility for the facility to allocate resources to manage their energy profile allowing a facility to operate their core business. The system operators have knowledge of local electricity sectors that continually monitor and update system protocols as regional markets change. The information is used to optimize the value realized by the ESaaS system while still meeting facility requirements.

### Environmental

For most ESaaS services, energy is stored during night time, off-peak hours when energy production is created from non-carbon emitting sources. The energy is then used to offset the required carbon emitting production during peak-times. The load shifting capability provided by ESaaS displaces heavy emitting generation requirements.

### Pricing

ESaaS contracts have a fixed monthly price over a contracted term. The fixed price is based on potential economic benefit and applicable programs in the region of deployment. The ESaaS contract price is always less than the economic value provided by the service to ensure the client retains a net positive value through the service.

## Energy Carrier

An energy carrier is a substance (energy form) or sometimes a phenomenon (energy system) that contains energy that can be later converted to other forms such as mechanical work or heat or to operate chemical or physical processes. Such carriers include springs, electrical batteries, capacitors, pressurized air, dammed water, hydrogen, petroleum, coal, wood, and natural gas. An energy carrier does not produce energy; it simply contains energy imbued by another system.

### Definition According to ISO 13600

According to ISO 13600, an energy carrier is either a substance or a phenomenon that can be used to produce mechanical work or heat or to operate chemical or physical processes. It is any system or substance that contains energy for conversion as usable energy later or somewhere else. This could be converted for use in, for example, an appliance or vehicle. Such carriers include springs, electrical batteries, capacitors, pressurized air, dammed water, hydrogen, petroleum, coal, wood, and natural gas.

ISO 13600 series (ISO 13600, ISO 13601 and ISO 13602) are intended to be used as tools to define, describe, analyse and compare technical energy systems (TES) at micro and macro levels:

- ISO 13600 (*Technical energy systems — Basic concepts*) covers basic definitions and terms needed to define and describe TESs in general and TESs of energyware supply and demand sectors in particular.

- ISO 13601 (*Technical energy systems — Structure for analysis — Energyware supply and demand sectors*) covers structures that shall be used to describe and analyse sub-sectors at the macro level of energyware supply and demand

- ISO 13602 (all parts) facilitates the description and analysis of any technical energy systems.

## Definition within the Field of Energetics

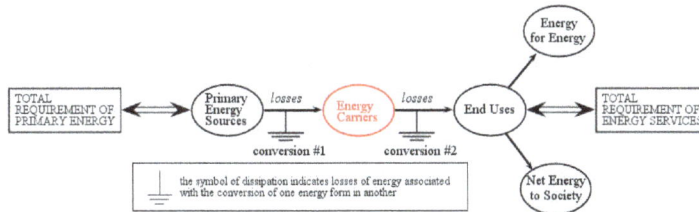

Energy carriers are produced by the energy sector using primary energy sources

In the field of energetics, an energy carrier corresponds only to an energy form (not an energy system) of energy input required by the various sectors of society to perform their functions. Energy carriers (EC) are produced by the energy sector using primary energy sources (PES). The distinction between "Energy Carriers" and "Primary Energy Sources" is extremely important since these two labels refer to energy forms of different quality so that they cannot be aggregated: 1 megajoule (MJ) of EC is not the same as 1 MJ of PES. Sunlight is a main form of primary energy, which can be transformed into plants and then into coal, oil and gas. Solar power and wind power are other derivatives of sunlight. Note that although coal, oil and natural gas are derived from sunlight, they are considered primary energy sources which are extracted from the earth (fossil fuels). Natural uranium is also a primary energy source extracted from the earth but does not come from the decomposition of organisms (mineral fuel).

## Leyden Jar

Early water-filled Leyden jar, consisting of a bottle with a metal spike through its stopper to make contact with the water.

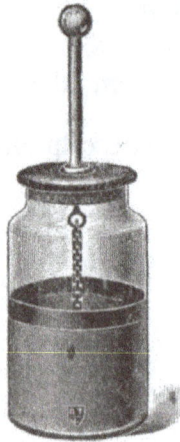

Later more common type using metal foil, 1919

A Leyden jar, or Leiden jar, is a device that "stores" static electricity between two electrodes on the inside and outside of a glass jar. A Leyden jar typically consists of a glass jar with metal foil cemented to the inside and the outside surfaces, and a metal terminal projecting vertically through the jar lid to make contact with the inner foil. It was the original form of a capacitor (originally known as a "condenser").

It was invented independently by German cleric Ewald Georg von Kleist on 11 October 1745 and by Dutch scientist Pieter van Musschenbroek of Leiden (Leyden) in 1745–1746. The invention was named after the city.

The Leyden jar was used to conduct many early experiments in electricity, and its discovery was of fundamental importance in the study of electrostatics. The Leyden jar was the first means of storing an electric charge which then could be discharged at the experimenter's will. Leyden jars are still used in education to demonstrate the principles of electrostatics.

## History

Discovery of the Leyden jar in Musschenbroek's lab. The static electricity produced by the rotating glass sphere electrostatic generator was conducted by the chain through the suspended bar to the water in the glass held by assistant Andreas Cuneus. A large charge accumulated in the water and an opposite charge in Cuneus' hand on the glass. When he touched the wire dipping in the water, he received a powerful shock.

A battery of four water-filled Leyden jars, Museum Boerhaave, Leiden

The Ancient Greeks already knew that pieces of amber could attract lightweight particles after being rubbed. The amber becomes electrified by triboelectric effect, mechanical separation of charge in a dielectric. The Greek word for amber is ("ēlektron") and is the origin of the word "electricity".

Around 1650, Otto von Guericke built a crude electrostatic generator: a sulphur ball that rotated on a shaft. When Guericke held his hand against the ball and turned the shaft quickly, a static electric charge built up. This experiment inspired the development of several forms of "friction machines", that greatly helped in the study of electricity.

The Leyden jar was effectively discovered independently by two parties: German deacon Ewald Georg von Kleist, who made the first discovery, and Dutch scientists Pieter van Musschenbroek and Andreas Cunaeus, who figured out how it worked only when held in the hand.

Despite its mundane and safe appearance, the Leyden jar is a high voltage device, and electrical energy collected within it from friction may be as high as 35,000 volts. The ball on the tip of the rod is a corona ball to prevent leakage of the energy into the air by point discharge.

## Von Kleist

Ewald Georg von Kleist (aka JG von Kleist) discovered the immense storage capability of the Leyden jar while working under a theory of electricity that saw electricity as a fluid, and hoped a glass jar filled with alcohol would "capture" this fluid. He was the deacon at the cathedral of Camin in Pomerania.

In 1744 von Kleist tried to accumulate electricity in a small medicine bottle filled with alcohol with a nail inserted in the cork. He was following up on an experiment developed by Georg Matthias Bose where electricity had been sent though water to set alcoholic spirits alight. He attempted to charge the bottle from a large prime conductor (invented by Bose) suspended above his friction machine.

Kleist was convinced that a substantial electric charge could be collected and held within the glass which he knew would provide an obstacle to the escape of the 'fluid'. He received a significant shock from the device when he accidentally touched the nail through the cork while still cradling the bottle in his other hand. He corresponded with a number of electrical experimenters, but didn't understand the significance of his conducting hand holding the bottle—and both he and his correspondents were loath to hold the device when told that the shock could throw them across the room. It took some time before Kleist's student associates at Leyden worked out that the hand provided an essential element.

## Musschenbroek and Cunaeus

For this reason, the Leyden Jar's invention was long credited to the Leyden physics professor, Pieter van Musschenbroek who also ran a family foundry which cast brass cannonettes, and a small business ('De Oosterse Lamp' -- "The Eastern Lamp") which made scientific and medical instruments for the new university courses in physics and for scientific gentlemen keen to establish their own 'cabinets' of curiosities and instruments.

Andreas Cuneaus appears to have been the first to receive communications from von Kleist about the storage capacity of the jar. He attempted to duplicate the experiment using a glass of beer, but couldn't make it work. He then worked with the Professor of Physics at Leyden University, and they eventually charged a jar of water with electricity only by holding it in the hand, rather than mounting it on an insulating resin block. Cuneaus and Musschenbroek also received severe shocks, and Musschenbroek communicated the experiment to Abbé Nollet, René Antoine Ferchault de Réaumur, and the wider French scientific community.

Musschenbroek's outlet in France for the sale of his company's 'cabinet' devices was the Abbé Nollet (who ran a similar business). Nollet then gave the electrical storage device the name "Leyden Jar" and promoted it as a special type of flask to his market of wealthy men with scientific curiosity. The "Kleistian jar" was therefore promoted as the *Leyden Jar* as having been discovered by Pieter van Musschenbroek and his assistant Andreas Cuneaus at the University of Leiden.

Daniel Gralath was the first to connect several jars in parallel to increase the total possible stored charge. The term "battery" was coined by Benjamin Franklin for these combinations, who likened it to a battery of cannon (cannons grouped in a common place). The term was later used for combinations of multiple electrochemical cells, the modern meaning of the term "battery". By the middle of the 19th century, the Leyden jar had become common enough for writers to assume their readers knew of and understood its basic operation.

Around the turn of the century it began to be widely used in spark-gap transmitters and medical electrotherapy equipment. By the early 20th century, improved dielectrics and the need to reduce their size and undesired inductance and resistance for use in the new technology of radio caused the Leyden jar to evolve into the modern compact form of capacitor.

## Design

A typical design consists of a glass jar with conducting tin foil coating the inner and outer surfaces. The foil coatings stop short of the mouth of the jar, to prevent the charge from arcing between the foils. A metal rod electrode projects through the stopper at the mouth of the jar, electrically connected by some means (usually a hanging chain) to the inner foil, to allow it to be charged. The jar is charged by an electrostatic generator, or other source of electric charge, connected to the inner electrode while the outer foil is grounded. The inner and outer surfaces of the jar store equal but opposite charges.

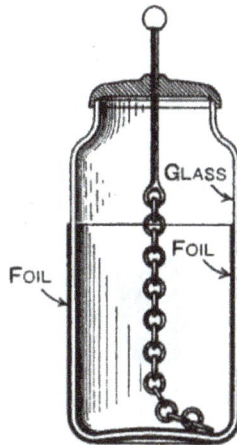

Leyden jar construction

The original form of the device was just a glass bottle partially filled with water, with a metal wire passing through a cork closing it. The role of the outer plate was provided by the hand of the experimenter. Soon it was found that it was better to coat the exterior of the jar with metal foil (Watson, 1746), leaving the (accidentally) impure water inside acting as a conductor, connected by a chain or wire to an external terminal, a sphere to avoid losses by corona discharge. Later the water inside was replaced with a second metal foil lining. Early experimenters found that the thinner the dielectric, the closer the plates, and the greater the surface, the greater the charge that could be stored at a given voltage.

Further developments in electrostatics revealed that the dielectric material was not essential, but increased the storage capability (capacitance) and prevented arcing between the plates. Two plates separated by a small distance also act as a capacitor, even in a vacuum.

Originally, the amount of capacitance was measured in number of 'jars' of a given size, or through the total coated area, assuming reasonably standard thickness and composition of the glass. A typical Leyden jar of one pint size has a capacitance of about 1 nF.

## Storage of the Charge

It was initially believed that the charge was stored in the water in early Leyden jars. In the 1700s American statesman and scientist Benjamin Franklin performed extensive investigations of both water-filled and foil Leyden jars, which led him to conclude that the charge was stored in the glass, not in the water. A popular experiment, due to Franklin, which seems to demonstrate this involves taking a jar apart after it has been charged and showing that little charge can be found on the metal plates, and therefore it must be in the dielectric. The first documented instance of this demonstration is in a 1749 letter by Franklin. Franklin designed a "dissectible" Leyden jar *(right)*, which was widely used in demonstrations. The jar is constructed out of a glass cup nested between two fairly snugly fitting metal cups. When the jar is charged with a high voltage and carefully dismantled, it is discovered that all the parts may be freely handled without discharging the jar. If the pieces are re-assembled, a large spark may still be obtained from it.

"Dissectible" Leyden jar, 1876

This demonstration appears to suggest that capacitors store their charge inside their dielectric. This theory was taught throughout the 1800s. However, this phenomenon is a special effect caused by the high voltage on the Leyden jar. In the dissectible Leyden jar, charge is transferred to the surface of the glass cup by corona discharge when the jar is disassembled; this is the source of the residual charge after the jar is reassembled. Handling the cup while disassembled does not provide enough contact to remove all the surface charge. Soda glass is hygroscopic and forms a partially conductive coating on its surface, which holds the charge. Addenbrook (1922) found that in a dissectible jar made of paraffin wax, or glass baked to remove moisture, the charge remained on the metal plates. Zeleny (1944) confirmed these results and observed the corona charge transfer. In capacitors generally, the charge is *not* stored in the dielectric, but on the inside surfaces of the plates, as can be observed from capacitors that can function with a vacuum between their plates.

## Residual Charge

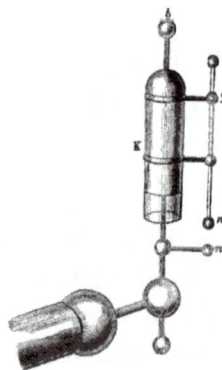

Measuring Leyden jar

If a charged Leyden jar is discharged by shorting the inner and outer coatings and left to sit for a few minutes, the jar will recover some of its previous charge, and a second spark can be obtained from it. Often this can be repeated, and a series of 4 or 5 sparks, decreasing in length, can be obtained at intervals. This effect is caused by dielectric absorption.

# Dry-seal Wiggins Gasholder

A dry-seal Wiggins gasholder is a device designed to hold gas.

A dry-seal gasholder can be designed to have a gross (geometric) volume ranging from 200 to 165,000 m³ (7,100 to 5,826,900 cu ft), whilst having a working pressure range between 15 and 150 millibars (1.5 and 15.0 kPa). The dry-seal gasholder is finished with an anti-corrosive treatment to counteract local climatic conditions and also any chemical attack from the stored medium. This anti-corrosive treatment is fully compatible with the sealing membrane and also the environment.

The dry seal gasholder has four major elements: the foundation; the main tank; the piston and the sealing membrane. Each of these elements can be divided into various sub-elements and associated accessories.

## Foundation

A concrete and hardcore base designed to withstand the weight of the steel gasholder structure constructed upon it and to withstand dynamic climatic conditions acting upon the gasholder etc.

## Main Tank

The main tank is designed to accommodate the design requirements laid down by the customer and climatic conditions There are three main sub-elements to the tank:

Tank bottom

The tank bottom forms a gas tight seal against the foundation and is "coned up" to facilitate drainage to the periphery. The bottom is covered with steel plates. The outer annular plates are butt welded against backing strips, whilst the infill plates are lap welded on the top side only. Welded to the bottom infill plates is the:

Piston support structure

When the piston is depressurised it rests on a steel framework which is welded to the bottom plates.

Tank shell

The shell of the tank is designed to accommodate the imposed loads and the general data supplied by the client. The shell is of butt-welded design and is gas tight for approximately 40% of its lower vertical height (known as the gas space) at which point the seal angle is located. The remaining upper 60% (known as the air space) of the shell has in it various apertures for access and ventilation. Attached to the shell are various accessories:

Staircase tower

For external access to the roof of the gasholder and also incorporates access to the inside of the gasholder via the shell access doors. A locked safety gate is usually located at the base of the staircase to prevent any unauthorised access to the gasholder.

Shell access doors

Doors located at pertinent points allowing access into the gasholder from the external staircase tower.

Shell vents

Allow air to be displaced from the inside of the gasholder as the piston rises.

Inlet nozzle

The connection nozzle allowing the stored gas to enter the gasholder from the supply gas main.

Outlet nozzle

For the export of the stored gas, this nozzle comes complete with an anti-vacuum grid to protect the sealing membrane during depressurisation. Depending on the operational process the inlet and outlet nozzles maybe a shared connection.

Shell drains

Allow condensates within the gasholder gas space to drain away in seal pots. The seal pots are designed to maintain the pressure with the gasholder.

Shell manways

Used for maintenance access into the gas space – only used whilst the gasholder is out of service.

Earthing bosses

To ensure that the gasholder is safe during electrical storms etc.

Volume relief pipes

Essential fail-safe system to protect the gasholder from over-pressurisation. Once actuated, by the piston fender, the volume relief valves allow the stored gas to escape to atmosphere at a safe height above the gasholder roof. As the volume relief valves open they actuate a limit switch.

Volume relief limit switches

Used to send signals to the control room to confirm the status of the volume relief valves.

Level weight system

A mechanical counter balance system to ensure that the pistons moments are kept in equilibrium. The level weights, which run up and down tracks located on the gasholder shell, also actuate limit switches to signal when the gasholder volume has reached pre-defined settings.

Level weight limit switches

Used to send signals to the control room to operate import and export valves etc.

Contents scale

On the gasholder shell is a painted scale displaying the volume of gas stored within the gasholder. An arrow painted on an adjacent level weight indicates the current status. Also painted on the scale is the location of the piston in relation to the shell access doors.

Seal angle

Welded to the inside of the shell this angular section is where the sealing membrane attaches to the shell.

Tank roof

The roof is designed to withstand the local climatic conditions and the possibilities of additional loads, such as snow and dust. The roof of the gasholder is of thrust rafter radial construction and has a covering of single sided lap welded steel plates. The roof has various accessories attached including:

Centre vent

Allows air to enter and exit the gasholder as the storage volume changes.

Roof vents

Small nozzle around the periphery used for the installation of the seal.

Roof manways

Allows access down to the piston fender when the gasholder is full.

Circumferential handrailing

Safety handrailing around the outside of the roof.

Radial walkway

For access from the staircase to the centre vent etc.

Volume relief valve actuators

Mechanical arms that operate the volume relief valves once the piston fender reaches a certain level.

Level weight pulley structures

Steel structures mounting the level weight rope pulleys and rope separators.

Load cell nozzles

For maintenance access to the load cell instrumentation used for volume recording purposes.

Radar nozzles

For maintenance access to the radar instrumentation used for volume recording purposes and piston level readings.

Roof interior lighting nozzles

For maintenance access to the gasholders interior lights.

## Piston

The gasholder piston moves up and down the inside of the shell as gas enters and exits the gasholder. The weight of the piston (less the weight of the level weights) produces the pressure at which the gasholder will operate. The piston is designed to apply an equally distributed weight to ensure that the piston remains level at all times. The piston made up of the following sub-elements:

Piston deck

The outer annular area is formed from butt welded steel plates resting on steel section rest blocks. Lap welded steel infill plates form a dome profile to withstand the gas pressure in the gas space beneath it. For higher pressure gasholders the infill plates are lap welded on both sides, whereas, low pressure gasholders are only welded on the top side. The fully welded piston deck forms a gas tight surface, which rests on the piston support structure when the gasholder is depressurised. The following ancillary items can be found on the piston deck:

Piston manway

Used for maintenance access below the piston into the gas space – only used whilst the gasholder is out of service.

Load cell chain receptacle

A receptacle for gathering up the load cell chains as the piston rises.

Piston seal angle

Welded to the outer top side of the annular plates, this angular section is where the sealing membrane attaches to the piston.

Level weight rope anchors

Equally spaced around the periphery of the piston deck are the connections to which the level weight ropes are fixed.

Piston fender

The fender is a steel frame structure that is fixed to the piston deck annular plates and acts as a support structure for the abutment plates. Access can be gained to the top of the piston fender from either the shell access doors or roof manways depending on the gasholder volume. Attached to the piston fender are the following items:

Piston walkway

> A platform around the top of the piston fender equipped with safety handrailing, used for inspection purposes.

Piston ladders

> Rung ladders complete with safety loops for access to the piston deck from the piston walkway.

Radar reflector plates

> Used to bounce the radar signal back to the radar instrument for volume indication recording and piston level readings.

Abutment plates

> Fixed to the outside of the piston fender to form a circumferential surface for the sealing membrane to roll against whilst the piston moves during operation.

Piston torsion ring

> Around the base of the piston fender is a torsion ring which helps keep the piston shape during pressurisation. Concrete ballast can be added to the torsion ring to increase the weight of the piston and subsequently be a cost-effective way to increase the pressure of the gasholder to the required level.

## Sealing Membrane

The seal of the gasholder is designed to operate in the conditions specified by the client and to suit the stored medium. The seal rolls from the shell to the abutment surface of the piston and vice versa providing the piston with a frictionless self-centering facility. During depressurisation the seal (Usually rubber) also provides a gas tight facility that protects the holder from vacuum damage by blocking the gas outlet nozzle. During commissioning of the gasholder the sealing membrane is set into an operating condition. This setting must be carried out every time the gasholder is depressurised, otherwise known as "popping" the seal.

## References

- "Power-to-X: entering the energy transition with Kopernikus" (Press release). Aachen, Germany: RWTH Aachen. 5 April 2016. Retrieved 2016-06-09

- Kvenvolden, Keith A. (2006). "Organic geochemistry – A retrospective of its first 70 years". Organic Geochemistry. 37: 1–11. doi:10.1016/j.orggeochem.2005.09.001

- "Stem: Energy Strorage-as-a-Service, Delivering Value to Customers and Utilities". Harvard Technology and Operations Management. Retrieved 2 January 2017

- Salpakari, Jyri; Mikkola, Jani; Lund, Peter D (2016). "Improved flexibility with large-scale variable renewable power in cities through optimal demand side management and power-to-heat conversion". Energy Conversion and Management. 126: 649–661. ISSN 0196-8904. doi:10.1016/j.enconman.2016.08.041

- Giampietro, Mario; Mayumi, Kozo (2009). The Biofuel Delusion: The Fallacy of Large Scale Agro-Biofuels Production. Earthscan, Taylor & Francis group. p. 336. ISBN 978-1-84407-681-9

- "Compressed Air Energy Storage (CAES)". energystorage.org. Energy Storage Association. Retrieved 13 February 2016

- Pagliaro, Mario; Konstandopoulos, Athanasios G (15 June 2012). Solar Hydrogen: Fuel of the Future. Cambridge, United Kingdom: RSC Publishing. ISBN 978-1-84973-195-9. doi:10.1039/9781849733175

- Beatty, Bill (1996). "Capacitor complaints". Science misconceptions in K-6 textbooks and popular culture. Science Hobbyist website. Retrieved 2010-06-12

- Zeleny, John (December 1944). "Observations and Experiments on Condensers with Removable Coats". Am. J. Phys. USA: AAPT. 12 (6): 329–339. Bibcode:1944AmJPh..12..329Z. doi:10.1119/1.1990632

# PERMISSIONS

# Index